V&R unipress

Aklilu Dalelo / Till Stellmacher

Faith-based Organisations in Ethiopia

The Contribution of the Kale Heywet Church
to Rural Schooling, Ecological Balance
and Food Security

With 22 figures

V&R unipress

Bonn University Press

Bibliographic information published by the Deutsche Nationalbibliothek

The Deutsche Nationalbibliothek lists this publication in the Deutsche Nationalbibliografie; detailed bibliographic data are available in the Internet at http://dnb.d-nb.de.

ISBN 978-3-8471-0011-9
ISBN 978-3-8470-0011-2 (E-Book)

**Publications of Bonn University Press
are published by V&R unipress GmbH.**

This book was published with financial support from the Evangelischer Entwicklungsdienst e.V.

© Copyright 2012 by V&R unipress GmbH, D-37079 Goettingen
All rights reserved, including those of translation into foreign languages. No part of this work may be reproduced or utilized in any form or by any means, electronic or mechanical, including photocopying, microfilm and recording, or by any information storage and retrieval system, without permission in writing from the publisher.
Printing and binding: CPI Buch Bücher.de GmbH, Birkach

Printed in Germany.

Inhalt

Acknowledgements 9

About the authors 11

List of Tables 13

List of Figures 15

List of Acronyms 17

1. Introduction 19
   1.1. Faith-based Organizations and the Genesis of Modern Education . 19
   1.2. Why Ecological Balance? 20
   1.3. Why Concern about Food Security? 23
   1.4. Rationale and Objectives of the Study 27
   1.5. Methodology 27
   1.6. Organisation of the Research Report 30

2. Ethiopia: Biophysical and Socio-economic Conditions 31
   2.1. The Biophysical Environment 31
   2.2. Socio-economic Environment 32
   2.3. Developments in the Educational Sector 35
   2.4. Food Insecurity in Ethiopia 37
       2.4.1. Causes of Food Insecurity 41
       2.4.2. Consequences of Food Insecurity and Famine 44
   2.5. The Ethiopian Food Security Strategy 46
   2.6. The Case Study Areas 48
       2.6.1. Kembata Tembaro zone and Alaba Liyu Wereda 48
       2.6.2. Adaa Liben Wereda 52

3. Faith-based Organisations and Development 55
   3.1. Perspectives on Faith and Development 55
   3.2. Factors Enhancing FBOs' Contributions 56
      3.2.1. A Clear and Accommodative Mission 56
      3.2.2. A Conducive International Atmosphere 59
   3.3. The Major Christian FBOs in Ethiopia 60
      3.3.1. Background 60
      3.3.2. The Early Missionaries 61
      3.3.3. Current Role of Evangelical Churches 62
   3.4. The Ethiopian Kale Heywet Church 64
      3.4.1. Background 64
      3.4.2. Educational Services of the EKHC 64
      3.4.3. Current Situation and Future Direction 66

4. Faith, Ecological Balance and Food Security: Survey Findings 69
   4.1. Interviewees' Profile 69
   4.2. Views on the Human Impact on the Environment 70
   4.3. Views on the Link between Environmental Protection and Food
        Security 71
   4.4. Views on Causes of Food Insecurity 73
   4.5. Views on Strengths and Special Contributions of FBOs 76
   4.6. Views on Weaknesses of FBOs 77
   4.7. Views on Evangelicals and Work Ethics 78
   4.8. Concluding Remarks 80

5. Case Study One: Empowering Schools to Address Ecological Balance . 83
   5.1. Background 83
   5.2. The Project Schools 84
   5.3. Project Outcomes and Impacts 85
      5.3.1. Preparation of a Handbook and Conducting In service
             Training 85
      5.3.2. Establishment of Environmental Clubs 87
      5.3.3. Development of School Nurseries 90
      5.3.4. Dissemination of Alternative Energy Technologies 90

6. Case Study Two: Improving Access to Education 95
   6.1. Background 95
   6.2. Key Objectives and Beneficiaries 96
   6.3. Major Achievements 96

7. Case Study Three: Help a Child	103
7.1. Background	103
7.2. EKHC Children's Services	106
7.3. The Kuriftu Children's Home	107
7.4. Community-Based Child Sponsorship	109
7.5. A Turning Point in the EKHC-ReK Partnership	111
7.6. Concluding Remarks	114
8. Case Study Four: Transforming Theological Schooling	115
8.1. Background	115
8.2. An Untapped Resource	116
8.3. Situation before the Integration Efforts	116
8.4. Integration of Development Education	117
8.5. Community Development at ETC	120
8.6. Concluding Remarks	122
9. Concluding Summary and Way Forward	125
9.1. Concluding Summary	125
9.2. The Way Forward	127
References	129
Appendix I	135

Acknowledgements

This study benefited from the support of many organisations and individuals. The idea to conduct research on the contributions of faith-based organisations to education and sustainable development with a focus on ecological balance and food security in Ethiopia was conceived when Dr. Aklilu Dalelo served as a technical advisor for the Ethiopian Kale Heywet Church (EKHC) Development Programme. The advisory service was made possible by the financial assistance of the Evangelische Entwicklungsdienst (EED) in Germany. Thanks and appreciation therefore go to the EKHC and EED. We would also like to thank Dr. Tesfaye Yacob, the then EKHC General Secretary, for his interest in the research and sustained encouragement throughout the entire process. Dr. Aklilu Dalelo and Dr. Till Stellmacher are grateful to Mr. Tefera Talore, Head of the Training Department at the EKHC, who supported the data gathering and processing. We are also greatly indebted to the key informants and others who gave information at different phases of the study.

The Center for Development Research (ZEF) at the University of Bonn allocated a variety of infrastructural and intellectual services and facilities to realise this work. Our heartfelt gratitude therefore goes to Professor Dr. Solvay Gerke, Director of the ZEF Department of Political and Cultural Change. The resource centre of the Oxford Centre for Mission Studies (OCMS), UK, has also been helpful. We would hence like to express our thanks to the staff of OCMS and Professor Deryke Belshaw, for their useful comments during the stay of Dr. Aklilu Delelo at OCMS. Likewise, the Human Needs and Global Resources Program (HNGR) of Wheaton College, USA, supported the field research by availing resources and office spaces. Our deepest gratitude goes to the staff of the HNGR Program and Department of Politics and International Relations. Professors Paul Robinson and Sandra Joireman made Dr. Aklilu Delelo's stay at Wheaton not only fruitful but also very enjoyable.

Two scholars reviewed the research report, Dr. Zemede Asfaw and Professor Ensermu Kelbessa, both from the Faculty of Science, Addis Ababa University. We would like to thank them for their critical and valuable comments.

Many institutions contributed financially to this study. The EED Scholarship Desk took responsibility for travel and research costs as well as costs related to review and publication. Our special thanks are due to Mrs Susanne Werner of the EED Scholarship Desk for doing much more than what her official responsibility required. The research within Ethiopia was mainly supported by the EKHC and Red een Kind (ReK), the Netherlands. Dr. Aklilu Delelo's stay at Wheaton College was sponsored by the HNGR Program, Wheaton College. We thank you all. Dr. Aklilu Dalelo has been greatly supported by the consistent encouragement of his wife, Tigist Tesfaye, and their children Elelta and Kranz. Dr. Till Stellmacher would like to thank his family for their partience and support throughout the writing of this book.

Dr. Aklilu Dalelo and Dr. Till Stellmacher
August 2012, Addis Ababa and Bonn

About the authors

Dr Aklilu Dalelo is an Assistant Professor of Geography and Environmental Education at Addis Ababa University, Ethiopia. He is a geographer by training with a particular interest in environmental and sustainability education. Dr Aklilu made an extensive research over the past fifteen years mainly in energy and environmental education; and published more than ten articles in reputable national and international journals. He also wrote a book entitled "Environment and Sustainability in Ethiopian Education System: A Longitudinal Analysis". Currently, he is an Alexander von Humboldt Research Fellow at the Institute of Environmental and Sustainability Communication, Faculty of Sustainability Sciences; University of Luneburg, Germany.
Email: akliludw@gmail.com; dalelowa@leuphana.de

Dr Till Stellmacher is a Senior Researcher at the Center for Development Research (ZEF), Germany. He studied Geography and Agricultural Sciences at the University of Bonn, Germany, and the University of Manchester, UK. For the last 10 years Dr Stellmacher worked extensively on natural resource management and conservation, governance, and the transformation of smallholder agriculture. His geographical focus is rural Ethiopia. Beyond he conducted research in Bangladesh, Burkina Faso, Nicaragua, India, the Ivory Coast and Tanzania. Dr Stellmacher serves as a lecturer at Addis Ababa University and the University of Bonn. Since 2011 he also works as a trainer for the GIZ.
E-mail: t.stellmacher@uni-bonn.de

List of Tables

Table 2.1: Population affected by food shortage: 1973/74 – 2000/01
Table 2.2: Average number of people affected over the last two-and-a-half decades
Table 2.3: Domestic production and food aid in Ethiopia
Table 2.4: Classification of food-insecure households in Ethiopia
Table 2.5: Total budget and food aid in Ethiopia
Table 3.1: Models that explain approaches to development and other religious activities
Table 3.2: Enrolment in non-state schools (1968)
Table 3.3: Distribution of Evangelical churches by region
Table 4.1: Participants' area of services
Table 4.2: Participants' level of education
Table 4.3: Participants' views about man's impact on nature
Table 4.4: Depth of analysis on the relationship between environmental protection and food security
Table 4.5: Causes of food insecurity
Table 4.6: Participants' views about aspects of food security
Table 4.7: Strengths and special contributions of FBOs
Table 4.8: Major weaknesses of FBOs
Table 5.1: Distribution of primary schools by *Wereda*, 2005
Table 5.2: Environmental clubs establishment and membership
Table 6.1: Education centre construction
Table 6.2: Enrolment
Table 6.3: Teachers' profile
Table 7.1: Educational level of the children at Kuriftu Home (as of 2003)
Table 7.2: Children graduated from colleges as of 2009
Table 7.3: Number and grade level of children supported through ReK in Babogaya
Table 7.4: Enrolment by region
Table 8.1: Number of directors and teachers trained and teaching materials prepared

List of Figures

Fig. 1.1: Location of management offices of case study projects
Fig. 2.1: Framework for analysis of famine causation and the appropriate responses
Fig. 2.2: Larger scale map of the study *Weredas*
Fig. 2.3: Enset plantation in Angacha Wereda
Fig. 2.4: House threatened by soil erosion
Fig. 2.5: Eucalypt plantation
Fig. 5.1: A teacher showing a tree planted in his school compound
Fig. 5.2: A solar home system being installed
Fig. 5.3: A solar cooker
Fig. 5.4: A solar drier
Fig. 5.5: An improved cooking stove called 'Mirt'
Fig. 6.1: A typical basic education centre
Fig. 6.2: Additional classrooms constructed fully by the community
Fig. 6.3: Children showed up even before the completion of the education centres
Fig. 6.4: Teachers' training
Fig. 7.1: Children in a village basic education centre, southern Ethiopia
Fig. 7.2: Some of the students at Kuriftu centre celebrating the graduation of their long-time friends
Fig. 7.3: Girls enrolled in one of the basic education centres in Alaba Wereda, SNNPR
Fig. 8.1: Community Development Programme students attend a class on improved stoves
Fig. 8.2: Community development students demonstrating the use of an improved cooking stove
Fig. 8.3: Community development students demonstrating the use of a bio-sand water filter

List of Acronyms

ADLI	Agricultural development-led industrialisation
ALW	Alaba Liyu *Wereda*
CaBCEP	Capacity Building and Community Empowerment Program
CBCS	Community-based Child Sponsorship
CBO	Community-based Organisation
CCCD	Childcare and Community Development
CCFC	Christian Children's Fund of Canada
CIDA	Canadian International Development Agency
CRDA	Christian Relief and Development Association
CSA	Central Statistical Agency
CSO	Civil Society Organisation
DICAC	Development and Inter-church Aid Commission
DPPC	Disaster Prevention and Preparedness Commission
DTRC	Demographic Training and Research Centre
ECA	Economic Commission for Africa
EEA	Ethiopian Economic Association
EECF	Ethiopian Evangelical Church's Fellowship
EECMY	Ethiopian Evangelical Church Mekaneyesus
EED	Evangelisher Entwicklungsdienst
EKHC	Ethiopian Kale Heywet Church
EOC	Ethiopian Orthodox Church
EOTC	Ethiopian Orthodox Tewahdo Church
EPRDF	Ethiopian People's Revolutionary Democratic Front
ETC	Evangelical Theological College
FAO	Food and Agricultural Organisation
FBO	Faith-based Organisation
FGM	Female Genital Mutilation
GDP	Gross Domestic Product
GMD	Gospel Ministry Department
HNGR	Human Needs and Global Resources
IDCoF	International Development Consulting Form
IDR	Institute of Development Research
IEC	International Evangelical Church

IPCC	Intergovernmental Panel on Climate Change
KNH	Kindernothilfe
KT	Kembata Tembaro
MoFED	Ministry of Finance and Economic Development
MOLSA	Ministry of Labour and Social Affairs
NCA	Norwegian Church AID
NCFS	New Coalition on Food Security
NGO	Nongovernmental Organisation
OCMS	Oxford Centre for Mission Studies
PASDEP	Plan for Accelerated and Sustained Development to End Poverty
ReK	Red een Kind
SHS	Solar Home System
SIM	Sudan Interior Mission (now renamed Serving in Mission)
SNNPR	Southern Nations, Nationalities and Peoples Region
SPSS	Statistical Package for Social Sciences
TTS	Teacher Training School
WCED	World Commission on Environment and Development
ZEF	Centre for Development Research, Bonn

1. Introduction

1.1. Faith-based Organizations and the Genesis of Modern Education

The history of education in Ethiopia is inextricably related to churches and mosques. The Ethiopian Coptic Orthodox Church developed "an elementary system of education, which served not only the needs of the Church itself, but also the cultural needs of society in general" (Markakis, 1974:143). Sirgiw Gelaw (2007) shares this view and further reinforces it:

> In the area of Education, the EOTC [Ethiopian Orthodox Tewahdo Church] served as the major centre of education from the time of its establishment until the beginning of the 20th century. The intellectuals of the EOTC had been dedicating their lives to serving the church and providing free education to its followers. They devoted their lives to teaching in the daytime and praying during the night in the service of their country and God. The existing literature, paintings, art, music, Qene, law and traditional medicines are all products of their educational endeavours (Sirgiw Gelaw, 2007:22).

Ethiopian Coptic Orthodox Church schools were originally created in Ethiopia to serve communities of believers, to instruct children in Christian principles and literature and to impart them with church rituals, prayers and hymns (Pankhurst, quoted in Hailu, 1975:145). This way, the Ethiopian Coptic Orthodox Church has had a great part in influencing and shaping what is believed today to be the Ethiopian culture. Before the revolution in 1974, the Ethiopian Coptic Church was running two kinds of schools. On the one hand was the modernised church school system based on a curriculum developed by the Ministry of Education, while on the other hand was a school system that strictly followed the traditional syllabus (Hailu, 1975:146).

The Missionary Societies

Beside the Ethiopian Coptic Orthodox Church, the Protestant Church also played a historically significant role in educational development in Ethiopia. Protestant missionary societies established and managed large numbers of schools throughout the country until the 1974 revolution. One such society, which played a remarkable role in the education and health sector, was the Sudan Interior Mission (SIM, now called Serving in Mission). Atkins (2003:np) reports that "SIM and EKHC [Ethiopian Kale Heywet Church] had the largest private school system in the country", as a result of which "the Ministry of Education asked SIM to be responsible for the distribution of government textbooks to all protestant schools" throughout the country.

In a document produced as part of the event to commemorate the 75th anniversary of the Ethiopian Kale Heywet Church (EKHC), Atkins recalls that "Dr. Peter Cotterell, SIM Bible School Coordinator, developed a system for teaching the Amharic language, which was so successful that it was adopted by the Ministry of Education for use in their adult literacy program". What is more, a Teacher Training School (TTS) was established in Wondo to address the increasing need for better educated Ethiopian teachers. The Wondo TTS was visited by Emperor Haile Selassi I in 1963. Atkins also shows that there was fierce opposition against mission education. Some government officials did not allow missionaries to establish schools out of the compounds. Mr. Gilen Cain, the then Director of the SIM in Ethiopia, "took the matter to Emperor Haile Selassie I, who ordered the Ministry of Education to allow all registered organisations to have schools throughout the country" (Atkins, 2003:np). The advance into secondary education was moving ahead well when Col. Mengistu Haile Mariam's communistic inspired government closed all the SIM stations in 1975 and all educational activities of SIM were overtaken by the congregations of the EKHC.

The above introduction shows that faith-based organisations made noteworthy contributions to the genesis and development of modern education in Ethiopia. This study aims to answer questions about their current contribution and role.

1.2. Why Ecological Balance?

The End of the 20th century saw a great deal of academic debates on the impact of human interventions on their natural environment. There is now a widespread understanding that quality and quantity of natural environments has deteriorated worldwide, mainly due to overuse and/or misuse by human beings. Climate change, loss of biodiversity, deforestation, and desertification are key challenges of our time. A report issued by the Intergovernmental Panel on

Climate Change (IPCC) in 2007, for example, asserts that "global increases in carbon dioxide concentrations are due primarily to fossil fuels use and land-use change" (IPCC, 2007:2).

An ecological balance
Among the major characteristics of ecosystems is their capacity to maintain a steady state, or equilibrium (balance). An ecological balance refers to a state of dynamic equilibrium within a community of organisms in which genetic species and ecosystem diversity remain relatively stable, subject to gradual changes through natural succession. Conservation of life maintains such a balance, for example through biogeochemical cycles, food chains, or population control (Purohit and Agrawal, 2005). Population explosions are seldom seen in natural ecosystems because of stabilising biotic and abiotic factors, a process often referred to as ecological or environmental resistance (Wright and Nebel, 2002:84). Ecosystems also tend to maintain their functioning through external disturbances. Wright and Nebel (2002:101) show that there are situations where "disturbances and shifting biotic relationships not only [...] have little detrimental effects on an ecosystem, but may actually contribute to its ongoing functioning". This is what ecologists refer to as ecological or environmental resilience.

Limits to ecological resilience
Golley (1999) argues that the capacity of ecological systems to withstand human interventions is not clearly understood. Wright and Nebel (2002:101) shows manifold limits of ecosystem resilience. Golley (1999:236) underscores that "[i]f we have vital needs and cannot appeal to a clear and concise measure of the capacity of ecosystems to withstand human demand, then the control of demand must come from within humans". The past decades show, however, a continuous increase of human interventions and impacts on ecological systems, thereby causing increasing ecological imbalances. An ecological imbalance is considered a common denominator underlying most problems associated with the ways humans use and manage their environment.

Humans as the principal culprits and victims of ecological imbalance
Ecosystems provide people with a variety of ecological functions and services[1] (see for example Anderson et al., 2006; Cunningham and Cunningham, 2008; Jialin et al., 2009). They provide fertile fields, pollinate crops and provide pure

---

[1] Ecosystem functions refer to the habitat, biological or system properties or processes of ecosystems, whereas ecosystem service refers to the life supporting products and services directly or indirectly obtained through the structures, processes and functions of the ecosystem (Jialin et al., 2009:542).

air to breathe and water to drink. Moreover, ecosystems transform and utilise parts of the waste produced by humans. Human impacts on the environment therefore create a boomerang effect in which humans increasingly become victims of their own making. Despite the technical changes that have taken place in many parts of the world over the last two centuries, "humans remain biological creatures who are undeniably part of ecological cycles" (Anderson et al., 2006:10).

Ecological functions and services remain absolutely necessary for human society in general and the rural poor in particular. People in more developed counties use proportionally more ecosystem services than those in less developed countries. However, livelihoods of the rural poor in less developed countries most directly depend on ecological services, e. g. to provide for subsistence agriculture, grazing, harvesting, hunting or fishing. Where there is limited access to infrastructure that provides safe drinking water, electricity or fuel, people particularly rely on natural services to meet their basic needs. One of the most basic human needs is food. Terrestrial and marine ecosystem services are one prerequisite to achieve food security in rural areas of less developed countries. However, increasing ecological imbalance contributes to pervasive poverty and food insecurity in these areas.

A desperate call
Sustainable development is currently high on the agenda of policy makers, scholars and media worldwide (see for example Anderson et al., 2006; Cunningham and Cunningham, 2008). The need for sustainable development has also been echoed by Faith-based Organisations (FBOs)[2] including the World Evangelical Fellowship Theological Commission. The Commission has emphatically underlined that "[w]e affirm the concept of sustainable development, as that which seeks to provide an environment that promotes a life of dignity and well-being compatible with the continuation and integrity of supporting ecosystems" (EFTC, 1999:350). A similar suggestion was made to "measure development by its ability to sustain healthy and dignified standards of living without excessive destruction or abuse of people and ecosystems" (Tsele, 2001:207). Bragg writes in his book "The Church in Response to Human Need" that unless "we seek development through 'gentle' technology that works with nature instead of abusing it, our grandchildren will live (if they can) in a much less hospitable world depleted of nonrenewable resources and choked by our own wastes" (Bragg, 1987:46).

---

2 In this work FBOs are understood to be formal entities established by people who share the same faith in order to achieve certain defined objectives.

1.3. Why Concern about Food Security?

Changes in the physical environment in general, and the currently increasing problem of climate change in particular, can have serious repercussions on the four dimensions of food security, namely food availability, food accessibility, food utilisation and food system stability (FAO, 2008). Climate change has, for instance, a direct impact on biophysical factors such as plant and animal growth, water cycles, biodiversity and nutrient cycling, and the ways in which these are managed through agricultural practices and land use for food production (FAO, 2008:11–12). Climate change can also affect physical/human capital – such as roads, storage and marketing infrastructure, houses, productive assets, electricity grids and human health – which indirectly changes the economic and socio-political factors that govern food access and utilisation and can threaten the stability of food systems.

Food security is only established "when all people, at all times, have physical and economic access to sufficient, safe and nutritious food to meet their dietary needs and food preferences for an active and healthy life" (FAO, 1996, quoted in Cavalcanti, 2005:152). Food sufficiency indicates enough food with the correct amount of calories, whereas access marks the ability of households to take command over these supplies through production, exchange or transfer. Food security thus includes the supply of food (food availability) via increasing the purchasing power of poor people, as well as the provision of social protection for the poor via safety nets or similar programmes (Tassew Woldehanna, 2004). Food availability can be increased via domestic supply (increased agricultural production), commercial import or food aid from abroad.

At this juncture, it is important to underline that the availability of food in a certain area, however, tells us very little about household food security. The distinction made between food shortage, food poverty and food deprivation seems to give a better insight into the problem of food insecurity at household level (DeRose et al., 1998). According to DeRose et al., hunger "is produced when need outstrips food availability, but the determinants of both need and availability are complex: they are controlled by forces both proximate to and quite remote from the individual they affect" (DeRose et al., 1998:2). In the case of food shortage, food supplies within some bounded region (such as a district) fall below the amount needed by the region's population. In the case of food poverty, a household will be unable to obtain enough food to meet the needs of its members. In the case of food deprivation, the nutrients consumed by an individual in a household fall below what he or she needs.

DeRose et al. (1998) further argue that the aforementioned three situations are causally linked, without offering a simple cause-effect pattern. One of the possible reasons for food deprivation is food poverty, and one of the reasons for food poverty is food shortage. The simplest equation says that if there is not enough food in the region, some households will not have enough; if there is not enough food in the household, some members will go hungry. However, some households in areas of food shortage are more than adequately provisioned, while some households in non-food shortage areas are not able to meet their members' needs. There are households even in the most economically advanced countries that cannot fully satisfy their food requirements. It is thus "evident that household food security is not assured by increased food supplies alone, but also by the pattern of income distribution and access to such supplies" (Fassil G. Kiros, 2005:4–5).

Ethiopia as a hunger hotspot
The 2005 assessment report on the world food security situation (published by the Committee on Food Security of Food and Agricultural Organisation) states that the number of countries facing serious food shortages throughout the world – so-called 'hunger hotspots' – stood at 36 as of March 2005, with 23 in Africa, seven in Asia/Near East, five in Latin America and one in Europe (FAO 2005, quoted in Cavalcanti, 2005:156). According to the report, a "particular concern is reserved for Eritrea, Sudan, Kenya, Cote d'Ivoire, Guyana, Haiti, Afghanistan, Iraq and Palestine". In view of the present study, three points from this report are worth emphasising. First, 64 per cent of the countries with severe food shortage are found in Africa. Second, three of the five neighbours of Ethiopia fall under the category of countries that cause particular concern. Third, Ethiopia has not been included in the list of countries that cause particular concern. Is Ethiopia therefore in a better position than neighbouring Eritrea, Sudan and Kenya? The answer seems 'No.'

Food insecurity in Ethiopia
Food poverty incidence in Ethiopia is about 50 per cent at national level, 37 per cent in urban areas and 52 per cent in rural areas (Workneh Negatu, 2008:1). For decades, the country has depended on food aid from abroad in an attempt to fill the deficit. Indeed, there were times when food aid amounted to one-quarter of the total food grains consumed in the country (Tassew Woldehanna, 2004:40). As a result, the name, Ethiopia, is to a large degree associated with food shortages and famine (see, for example, Box 1.1).

Box 1.1: Ethiopia's food aid addiction

> Like a patient addicted to painkillers, Ethiopia seems hooked on aid. For most of the past three decades, it has survived on millions of tonnes of donated food and millions of dollars in cash. It has received more emergency support than any other African nation in that time. Its population is increasing by 2 m every year, yet over the past 10 years its net agricultural production has steadily declined. Even in good years, some 5 m people need food aid just to survive. Ethiopia is so poor that it takes one bad rainy season to tip millions more into crisis. But like any addict, Ethiopia seems to have taken the first step towards recovery by recognizing that there is a problem.
> Source: Greste, 2006

Food insecurity in Ethiopia is predominantly chronic in nature. There are households that cannot meet their food needs in any given year, regardless of climatic variables or other external and internal shocks. Chronically food-insecure households in the country include mainly those who are landless/land-scarce, oxen-less, pastoral, female headed, elderly, disabled and sick, new settlers, non-agricultural poor households and low-income urban households (Workneh Negatu, 2008:1). It should also be noted here that decades of food aid distribution have not substantially improved their nutritional or economic status.

Faith-based organizations and food insecurity in Ethiopia
The last three decades have seen continuous attempts by international governmental and NGOs to fight the problem of food insecurity in Ethiopia. Ethiopian and international FBOs have also been playing their part to help breaking the vicious cycle of food insecurity, environmental degradation and poverty. Most FBOs in Ethiopia have a long tradition of fighting food insecurity, mostly through their social or wholistic services (ministries)[3], and consider such activities an integral part of their mission. Some scholars see the work of FBOs in antagonism to capitalism in the sense that "without religion as its base, development will be reduced to an appendage of capitalist ideology and, therefore, will not offer much to the poor in Africa" (Tsele, 2001:205).

---

3 Most Evangelical denominations in Ethiopia are composed of local churches or congregations. Each congregation has services divided into what they call 'ministries' (examples include the Evangelism ministry, the children ministry, the women ministry, the social ministry, and the development ministry).

The role of Ethiopian FBOs during the famines in the 1970s and 1980s has been well documented (see details in Chapter Four). Grassroots initiatives related mainly to relief and rehabilitation activities were, for instance, conjointly directed by church organisations and regional drought committees. The churches (mostly Orthodox, Lutheran, Baptist, and Catholic) were "coordinated in their relief activities by the Christian Relief and Development Organisation" (Webb et al., 1992:95). Many FBOs served as local grassroots 'outlets' for international donors including philanthropic organizations (e.g. Band Aid) using their structural advantage due to their extended and multiple human, technical and institutional capacities at grassroots level.

Diesen and Walker (1999) stress the difficultly to describe the role of NGO activities in Ethiopia mainly due to the paucity of systematic studies conducted in the country. Belshaw (2006) underscored the need for empirical research on FBO's local work in Ethiopia. With regard to such a need, Fassil highlights that it is "never too late to begin to build the valuable tradition of learning from the past and instituting systems for distilling lessons of experience and for their retrieval to serve as the basis for planning of future development" (Fassil G. Kiros, 2005:172). Belshaw also stresses that "[s]ince a statistically reliable survey of both coverage and quality of Christian development impacts is not available, this deserves the highest priority for research attention in the context of growing opportunities and responsibilities available for the church to renew its historical pro-poor wholistic mission" (Belshaw, 2002:140). Such data, according to the above cited writers, can help better discuss the contributions of FBO activities in Ethiopia and encourage them to take a critical look at their own work (Abiy Hailu, 2004).

Amazingly little empirical research has been conducted on local work of FBOs in Ethiopia so far. Goyder and Wiegboldus (2006) did a study on emergency relief programmes implemented by four FBOs.[4] In 2003 Aklilu Dalelo published a study titled "The Church and Socio-economic Transformation" which basically focuses on the EKHC Development Programme.

---

4 Goyder and Wiegboldus (2006:3–4) indicate that emergency relief programmes implemented by four FBOs in Ethiopia have been effective in helping to reduce short-term food access of target communities, and have mostly been efficient in using funds to provide the maximum benefit for target communities given the chosen intervention activities and available capacity. In terms of impact and sustainability, however, the programmes did not have a direct effect.

1.4. Rationale and Objectives of the Study

This study addresses the shortage of research on the role of FBOs in Ethiopia in general and the role of the EKHC in education and sustainable development in particular and aims to contribute to the current debate within EKHC, in other FBOs, among development practitioners and scholars on these issues. The study was motivated by a desire to find out the contributions of FBOs in Ethiopia in their efforts to improve education and sustainable development with a focus on ecological balance and food security. FBOs in general have an enormous potential to reach the poor living in the remotest rural parts of less developed countries where government structures and facilities are often weak or even nonexistent. The Ethiopian Kale Heywet Church alone has more than 7,000 local churches spread throughout the country. Imagine the a potential of establishing 7,000 primary schools by mobilising these local churches. If each of the schools would enrol about 200 children, a total of 1.4 million children could get access to primary formal education.

The EKHC has been engaged for decades now in activities directly and indirectly related to educational development and ecological balance. However, no comprehensive study has been conducted so far to determine the outcomes and impacts of their interventions. The principal aim of the present study is hence to assess the potential and actual impacts of interventions undertaken by the EKHC with regard to educational development and ecological balance. Food security will be as a cross-cutting issue.

The study seeks to achieve the following specific objectives:
– to empirically assess the local activities and achievements of the EKHC in Ethiopia with regard to improving education and sustainable development with a focus on ecological balance and food security, and
– to discuss, based on these findings, measures that could enhance local capacities and effectiveness of FBOs in Ethiopia.

1.5. Methodology

The work follows cross-sectional and case study approaches. A triangulation of methods and sources has been used to diversify and validate sources of information. Both quantitative and qualitative methods were used, as described in the following section.

A. Document analysis

Secondary literature and internal documents of EKHC have been analysed to generate a current and comprehensive understanding on the work of FBOs in Ethiopia. Many organizations helped with providing secondary literature, namely the Center for Development Research (ZEF), Germany; the Oxford Centre for Mission Studies (OCMS), UK; the Human Needs and Global Resources (HNGR) Program, Wheaton College, USA; the Institute of Development Research (IDR), Ethiopia; and the Christian Relief and Development Association (CRDA), Ethiopia.

B. Survey with EKHC decision makers

A total of 206 EKHC decision makers from throughout Ethiopia were surveyed. In total 52 out of the 83 EKHC districts are represented. Given the vast geographic area of Ethiopia, this is quite a significant representation. Most of the participants were surveyed before or after regular meetings or training sessions. In some cases, the questionnaires were circulated by change agents (people hired to facilitate the process of leadership and ministry[5] transformation). The decision makers' survey was conducted from February to June 2007.

A comprehensive questionnaire (see Appendix I) was used for the survey. It has four sections, the first of which deals with general information about the participants (their sex, place of work, type of work / ministry, level of education). The second section is aimed at examining their knowledge of and views about the relationship between environmental protection and food security; causes of food insecurity; contributions of faith and religion[6] to socio-economic development; strengths and weaknesses of faith-based organisations and the state of the relationship between Ethiopian Evangelical churches and the Ethiopian government. The third section is meant to gather more specific information on practical activities at the grassroots level. Included in this section are issues such as the key values local churches want to inculcate into their followers; types of social / development services / ministries undertaken by local churches; strategies to obtain required resources; the state of advocacy work and whether it is

---

5 The term 'ministry' in this study refers to services members give in their local churches such as preaching, teaching, social services, development work, etc. Some local churches have more than fifteen such ministries.
6 The term 'faith' in this context refers to a trusting belief in a Supreme Being. In Christianity faith is based on the work and teaching of Jesus Christ. The term 'religion' is considered as a set of ideas, values, experiences and attitudes. People who share these ideas, values, experiences and attitudes come together to form a religious group.

easy or difficult to carry out development work in their areas. The last part of the questionnaire has an attitude scale composed of 37 statements. The scale is meant to measure the tendency of EKHC leaders and staff working in the development programmes regarding the relationship between spiritual and social/development activities carried out by their particular church. The scale was designed based mainly on models that show how FBOs deploy faith through social or political engagement (Geest, 1993; Clark, 2005). The data of the survey was processed using SPSS.

The draft version of the questionnaire was pilot tested. People with a similar profile to those who would complete the final version of the questionnaire were first asked to complete the questionnaire and then to comment on its effectiveness. Twelve experts and practitioners working in community development were then asked to examine critically the quality of the questionnaire.

C. Programme / project case studies

Four EKHC programmes/projects related to educational development and ecological balance were assessed on case study bases (Fig. 1.1). These are:
1. a school capacity building project, which operates in the SNNPR region with its management office located in Awassa town,
2. an alternative basic education project, which operates in SNNPR and Oromiya regions with its management office located in Alaba town,
3. the Kuriftu children's care project, which operates in the Oromiya region with its management office located in Debrezeit town, and
4. a project on integration of environmental and development education into the curricula for bible schools and theological colleges. This project is implemented in different parts of Ethiopia and coordinated by the EKHC Training Department, whose office is located in Addis Ababa.

A large number of project proposals, progress reports, monitoring and evaluation reports, correspondences with donors and government offices, etc. were studied in order to examine the projects' activities, achievements and limitations. Additional field surveys were conducted in the programme/project areas between February and June 2007. During these field surveys, focus group discussions were conducted with a total of 70 people. The groups of people participated at the focus group discussions, namely:
– community and religious leaders,
– school administrators and teachers, and
– beneficiaries of the different projects.

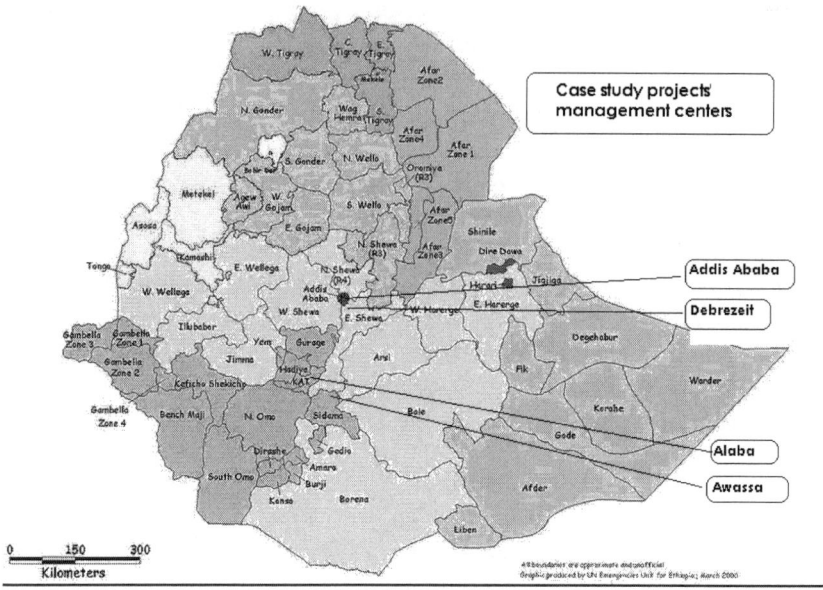

Fig. 1.1: Location of management offices of case study projects

1.6. Organisation of the Research Report

The report is divided into nine chapters. Chapter One introduces the topic and the methodological approaches, whereas Chapter Two discusses the biophysical and socio-economic conditions in Ethiopia. Chapter Three reviews the work of FBOs in Ethiopia. Chapter Four presents the empirical results of the survey with EKHC decision makers. Chapters Five to Eight present the four case study projects. The last chapter presents conclusions and the way forward.

2. Ethiopia: Biophysical and Socio-economic Conditions

2.1. The Biophysical Environment

Physiographic features
Ethiopia is among the largest countries in Africa, both in terms of geographic area and population. The land area covers 1,127,127 km^2 (MoFED, 2002). Ethiopia is of great geographic diversity with high and rugged mountains, plateaus and deep gorges incised by river valleys and rolling plains. Altitude varies from as low as 120 m below sea level at the Dallol depression to as high as 4,620 m above sea level in the Semien Mountains at Ras Dashen Peak. The highlands, with altitudes ranging from 1500 m to 3500 m, accommodate 88 per cent of the human population, 75 per cent of the livestock population and 95 per cent of the total cultivated land (EEA, 2005). In total, the Ethiopian highlands account for about half of all the highlands in Africa (Hurni, 1988, quoted in EEA, 2005).

Climatic conditions
Climatic conditions determine land use patterns, agricultural productivity and food security. With few exceptions, most parts of the western, central, southern and eastern highlands receive high and stable rainfall. Places such as Kaffa, Illubabor, Sidamo, Arsi-Bale highlands, parts of East Gojjam and parts of the Gamo highlands are known to receive mean annual rainfall of 2000 mm. In the north-eastern highlands and south-eastern lowlands, rainfall is generally much lower and more variable. On the whole, Ethiopia receives one of the highest levels of rainfall in Eastern Africa. The national data of rainfall documented over four decades indicates that, on average, the country receives rainfall above 1000 mm annually. However, the amount fluctuates from year to year. This fluctuation has been very frequent in recent years, mainly after the mid 1980s, implying an increasingly serious challenge to agriculture which is predominantly dependent on rainfall. Drought has been frequent as a result of late onset, abnormal distribution and early cessation of rainfall. Such a trend in the distribution of rainfall has serious implications for measures used to improve food security.

Primary forest cover

It has been a common practice to cite the alarming rate at which the primary forest of Ethiopia has been diminished over the last hundred years or so (Stellmacher, 2007a). Although there are some disagreements on the tempo and causes of deforestation that has taken place in the country, the diverse reports seem to agree that the remaining primary forest cover of the country covers not more than 3 per cent of the countries' area (Stellmacher, 2007b:519). In 2000, EarthTrends estimated that Ethiopia had 4,344 million hectares of primary forest area, which was 4 per cent (Gatzweiler, 2007). If the current rate of deforestation continued unabated, the report stated, it was feared that the "country [would] lose its last high forest tree within about 27 years" (Gatzweiler, 2007:4). There are strong linkages between deforestation and food insecurity. Smallholder farmers which are exposed to food insecurity have low time preference rates which often forces them to convert primary forests into agricultural land (Gatzweiler, 2007).

2.2. Socio-economic Environment

Demographic characteristics

Ethiopia's total population was estimated 73.9 million in 2007, the second largest in Sub-Saharan Africa. Out of the total population of the country, only 16.2 per cent live in urban areas. According to the Central Statistical Authority (CSA) projection (medium variant), the total population is estimated to reach over 130 million by 2025. The gross population density in 2000 was 59 persons per km^2, with a marked concentration in the highlands. Nearly half of the population live at altitudes of 2,200 m above sea level or higher, while only 11 per cent live at altitudes below 1,400 m. The remaining 40 per cent live in areas with an altitude ranging from 1,400 m to 2,200 m above sea level (Assefa Hailemariam, 2003).

The Ethiopian population is young. The median age is less than 18 years. Children under the age of 15 years constitute about 44 per cent, those aged 60 years and over make up less than 5 per cent, while the working age population constitutes about 52 per cent. Fertility increased in the 1980s and early 1990s from about 5.2 children per woman in 1970 to 7.5 in 1984, and further to 7.7 children per woman in 1990. In 1994, however, this figure began to decline, and in 2000 the total fertility rate was 5.9 children per woman (Assefa Hailemariam, 2003).

Dominant Christian denominations in Ethiopia

The most recent census reveals that the Ethiopian Coptic Orthodox Church accounts for 43.5 per cent of the total population, followed by Islam with 33.9 per cent (Population Census Commission, 2008:96). The Ethiopian Coptic Orthodox Church is numerically the largest of the five non-Chalcedonian Eastern Churches:

the Coptic, the Ethiopian, the Syrian, the Indian, and the Armenian (Getnet Tamene, 1998). The form of Christianity, which the Ethiopian Orthodox Church represents, dates back at least to the fourth century AD and is still exerting a powerful influence on the lives of millions. Though the Church received the spiritual and theological traditions of the Orthodox Church from its earliest days, she has been indigenized and has made the heritage which is her own and even developed it in a unique way against the cultural and social background of Ethiopia (Getnet Tamene, 1998). The Protestant and Evangelical churches make up the second most important Christian denomination in the country, accounting for 18.6 per cent of the total population, whereas Catholic churches account for only 0.7 per cent (Population Census Commission, 2008:96).

Recent socio-political developments
The last four decades have seen tremendous dynamics in the political landscape of the country, which have resulted in frequent changes in the social and economic structures. In 1974, the Emperor Haile Selassie was overthrown by a military government, which became known as the Dergue (Stellmacher, 2007b). During the 1980s, civil war intensified, especially in the north of the country, and caused enormous hardships. In general, Ethiopia remained a highly centralised country both during the Emperor's regime and the Dergue's tenure.

The Dergue were brought down in 1991 by a rebel group from Tigray, since when a number of reforms have been implemented including the decentralisation of decision making to the newly established regions. Currently, the country has nine administrative regions and two city administrations. These are Tigray, Afar, Amhara, Oromiya, Somali, Benishangul-Gumuz, Southern Nations, Nationalities and Peoples Region (SNNPR), Gambela, Harari, Addis Ababa and Dire Dawa. Four of these regions (Tigray, Amhara, Oromiya and SNNP) contain more than 90 per cent of the rural population (Dercon, 1999).

By 1988, the worsening economic conditions and pressure for change forced the then government to make modest economic reforms (Dercon, 1999). Early measures included the freeze on land reform, the abolition of the quota system and the removal of some of the restrictions on the movement of food across regions. After the fall of the Dergue, economic reforms continued. In 1992, a large devaluation took place and gradually many restrictions on the private and banking sectors were removed. Substantial international trade liberalisation was also implemented. In addition, input marketing was liberalised and extension activities reformed (see details on the Ethiopian economy in the succeeding section).

There is a longstanding debate as to the effect of land tenure on food security. The more dominant view says that state ownership of land in Ethiopia has negative effects on food security by stifling farmers' initiatives to adopt productivity-enhancing land improvement practices (IDCoF, 2002). Others dismiss such a conclusion as unfounded. Hussien Jemma (2001) argues that the current advocacy of privatising rural land in Ethiopia is "invalid for the simple reason that it failed to produce sufficient empirical evidence, which would otherwise show that it is imperative to replace existing public ownership with private tenure" (Hussien Jemma, 2001:73). According to Hussien's study, farmers from the Gedeo zone of SNNPR reported that they "have never felt insecure of their holding ever since the Land Reform Proclamation of the mid-1970s. No one has threaded or attempted either to redistribute land or displace them from their possessions. Redistribution is done only within a household to accommodate the young members that seek farmland" (Hussien Jemma, 2001:55).

In relation to forest resources, the 1994 Proclamation (No. 94/1994) made a distinction between the public and private ownership of forests, declared natural forests as state-owned and allowed planted forests to be owned privately. This proclamation also prohibits using or harvesting trees, settling, grazing, hunting or keeping beehives in the state forest. More recently, more participatory approaches to forest ownership and management (co-management approaches, participatory forest management) have been introduced with the help of international donor agencies. According to the new approach, forest user groups have been established and exclusive rights for forest use granted to the members of the group (Gatzweiler, 2007; Stellmacher, 2007a; Stellmacher and Nolten, 2010).

The Ethiopian economy in brief
The Ethiopian economy depends heavily on the agricultural sector, which contributes 55 per cent of the GDP of the country (ICDoF, 2002; EEA, 2005). The agricultural sector is dominated by subsistence farmers who produce largely for their own consumption and contribute very little to the market. In the plough-based cereal system, *teff*, wheat, maize and sorghum are grown as staples, mixed with livestock keeping. The average land holding is about 2 ha per household in these areas, and population density is about 90 persons per km^2 (FAO, 1986; quoted in EEA, 2005). In contrast, the hoe-based coffee-*enset*-livestock sub-system is characterised by high population density (200–350 persons per km^2) with an average holding of 0.5 ha (for a family of 8 persons) where intensive mixed faming is practiced. Some areas such as Kedida Gamella *Wereda* in the Kembata Tembaro zone are known to have a population density that exceeds 500 persons per km^2.

Performance of the Ethiopian agriculture
The performance of the Ethiopian agriculture sector is low by any standard. Land and labour productivity are among the lowest in the world. Between 1992/93 and 2002/03, agricultural growth averaged about 1.5 per cent per year, with

sharp variations in between, increasing up to 15 per cent in a particularly good year and declining by as much as 12 per cent in a drought year like 2002/03 (ICDoF, 2002; EEA, 2005). During the five-year period covering 1999/2000 to 2003/2004, earnings per agricultural person and farm household remained largely constant (EEA, 2005).

Agriculture's poor performance is associated highly with frequent droughts. At present, drought is considered the single most important climate-related natural hazard in Ethiopia (Abebe Tadege, 2008). Agriculture is one of the most vulnerable sectors to climate variability and change. As rain-fed agriculture is widely practiced in the country, variations in agricultural production (yield) closely follow variations in rainfall. Despite pressing needs for irrigation, Ethiopia irrigates only 3 – 5 per cent of its farmland from the potential 3.7 million ha of irrigable land (EEA, 2005).

Information on per capita income over the past four decades also indicates that the total performance of the Ethiopian agriculture sector is deteriorating. From 1953 to 1995, the per capita income in agriculture declined by over 45 per cent; researchers hence strongly argue for transforming the sector. It is regrettable that "no fundamental change of a desired magnitude could be brought about after four to five decades of agricultural research and extension programme activities" (EEA, 2005:147 – 148).

Government revenue from agricultural income tax has varied in recent years, excluding the 2002/03 drought year, between 100 and 138 million Birr, which is only 10 to 13 Birr per agricultural household (EEA, 2005). Agricultural export tax has been removed since 2002 to encourage export trade. According to EEA (2005), agriculture's contribution to government direct revenue in the past years has been on average only 3 per cent, which is too little and shows how the sector is weak in terms of surplus generation. Indirect surplus extraction from agriculture in terms of the supply of cheap food is also low.

2.3. Developments in the Educational Sector

The development of the education sector in Ethiopia is still in its infancy. On the eve of the ongoing educational reform process, which began in 1994 (following the endorsement of the New Education and Training Policy), enrolment in primary education stood at about 2.81 million (Ethiopian National Agency for UNESCO, 2001). This included over-age pupils, which amounted to 34 per cent of the school-age population at the time. Likewise, the enrolment ratio at secondary level stood at about 15 per cent and at the third level 1 per cent. Compared to African countries, Ethiopia's enrolment ratios have fared among the lowest in primary education. Similarly, enrolment at all levels of education is male biased, the tertiary level being the worst in this regard.

One should also emphasise here that the last couple of years have seen significant improvements in enrolment at all levels. It is believed that consecutive Education Sector Development Programmes (ESDPs) have contributed a great deal to this success (MoFED, 2006). The ESDP is a comprehensive intervention package developed by the Ethiopian government to mobilise national and international efforts to boost the performance of the system, in particular the primary education sub-sector. The government launched a twenty-year education sector indicative plan in a bid to implement the 1994 National Education and Training Policy. The first, second and third five-year plans (ESDP I, ESDP II and ESDP III) had already been implemented. During the ESDP I and ESDP II implementation period, noteworthy progress was made with regard to access and equity. ESDP III, on the other hand, was aimed at achieving the MDGs and meeting the objectives of the National Development Plan by supplying a qualified and fully trained workforce with the necessary quantity and quality at all levels.

The specific objectives of the ESDP III include (MoFED, 2006):
– ensuring education and training quality and relevance,
– lowering educational inefficiency,
– preventing HIV / AIDS,
– increasing participation in education and training and ensuring equity, and
– increasing the participation of stakeholders.

Furthermore, specific strategies had been designed to achieve the aforementioned objectives of ESDP III (MoFED, 2006). These included:
– strengthening community and NGO participation in the sector,
– using alternative methods to implement the programme in low cost school construction (these include the construction of low-cost classrooms and using them for first-cycle primary education),
– expanding primary education coverage, increasing the role of non-formal education and other alternatives,
– designing and implementing policies that strengthen communities' participation in the administration and financing of schools,
– building the capacity of institutions, providing training for school managers, developing and implementing logistic support systems,
– giving additional responsibility for *woredas* / districts to administer primary and secondary schools,
– improving ethical values of teachers beyond academic qualifications,
– providing support to increase women's educational participation and to help children that have special educational needs.

2.4. Food Insecurity in Ethiopia

Famine and food insecurity are by no means new events in Ethiopia. Nevertheless, the country has experienced increasingly more intensified problems concerning food shortages over the past four decades. For the period between 1973/74 to 2000/01, it is estimated that more than two million people per year (except in 1978/79) face food shortages of different magnitudes (Table 2.1). An attempt has been made to show the temporal pattern of food shortage by calculating the difference between the consecutive half-decades (the latter half-decade minus the earlier). The computation was made based on the same data indicated in Table 2.1. With four to seven million people affected every year, the first half of the 1990s stood first (Table 2.2), whereas the second half of the 1970s saw fewer numbers of people affected. There is, however, no traceable general trend in the number of people affected by food shortages.

Table 2.1: Population affected by food shortages: 1973/74 – 2000/01

Half-decade	Year	People affected (million)	Half-decade	Year	People affected (million)
	1973/73	3.0		1987/88	2.1
	1974/75	2.7	III	1988/89	No Info.
	1975/76	No Info.		1989/90	3.4
	1976/77	No Info.		1990/91	7.2
I	1977/78	2.7		1991/92	7.9
	1978/79	1.0	IV	1992/93	5.0
	1979/80	3.7		1993/94	6.7
	1980/81	3.3		1994/95	4.0
	1981/82	4.2		1995/96	2.8
II	1982/83	4.0	V	1996/97	3.4
	1983/84	5.0		1997/98	4.1
	1984/85	7.9		1998/99	5.4
	1985/86	6.9		1999/00	2.6
	1986/87	2.5		2000/01	7.7

Source: Based on data from DPPC (Quoted in Fassil G. Kiros, 2005:78)

Table 2.2: Average number of people affected over the last two-and-a-half decades

Half-decade	People affected per year on average (millions)	Difference between consecutive half-decades
I. Second half of 70s	2.47	–
II. First half of 80s	4.88	+2.41
III. Second half of 80s	3.73	-1.15
IV. First half of 90s	6.16	+2.43
V. Second half of 90s	3.66	-2.50

Source: Computed by the authors based on data indicated in Table 2.2

A more recent study shows that an average of about six million people were food insecure between 1996 and 2006 (EEA, 2008). The lowest number was known to be 2.69 million in 1996, while the highest was 12.2 million in 2003.

Excessive dependence on food aid
It is widely believed that more than half of Ethiopia's population has no access to the minimum nutritional requirement of 2100 kcal/person/day (IDCoF, 2002). Food aid from abroad plays a significant role in reducing the deficit. Ethiopia, according to EEA, has been a recipient of "food aid and other humanitarian assistance over many decades, to such an extent that emergency relief has become institutionalized within governmental structures and donor agency country programmes" (EEA, 2008:86). Indeed, there were times when food aid amounted to one-quarter of the total food grain production in the country (see Table 2.3*)*. Table 2.3 shows that food aid contributed for 26 and 17 per cent of the total food demand of the country in 1985 and 1986, respectively. For almost half of the time (eight out of the seventeen years), food aid has covered more than ten per cent of the national food demand.

Table 2.3: Domestic production and food aid in Ethiopia

Year	Aid (000's tonnes)	Production (000's tonnes)	Proportion of aid
1985	1,272	4,855	26.2
1986	926	5,404	17.1
1987	277	6,684	4.1
1988	1,096	6,902	15.9
1989	461	6,676	6.9
1990	657	6,579	10.0
1991	925	7,078	12.0
1992	840	7,055	11.9
1993	5,198	7,619	6.8
1994	980	6,945	14.1
1995	683	7,493	9.1
1996	334	10,328	3.2
1997	428	10,167	4.2
1998	615	8,036	7.7
1999	776	8,552	9.1
2000	1,380	8,890	15.5
2001	639	10,616	6.0

Source: Tassew Woldehanna, 2004:40

A largely chronic problem

Food insecurity in Ethiopia is predominantly chronic in nature, as many households cannot meet their food needs in any given year, regardless of climatic vagaries or other external shocks. Fassil G. Kiros (2005:16–17) describes chronic food shortage as a condition which inflicts damage, slowly sapping human energy and depleting the capacity to produce – and thereby increasing vulnerability to famine. Chronic food shortages, over extended periods of time due to malnutrition, ill-health and declining agricultural productivity, also hamper long-term development. Attempts have also been made to classify households in Ethiopia as living under chronic or transitory food insecurity (Table 2.4).

Table 2.4: Classification of food-insecure households in Ethiopia

	Rural	Urban	Others
Chronic	– Resource-poor households – Landless or land-scarce – Ox-less – Poor pastoralists – Female-headed households – Elderly – Disabled – Poor non-agricultural households – New settlers	– Low income households employed in the informal sector – Groups outside the labour market: elderly, disabled, some female-headed households	– Refugees – Displaced people – Ex-soldiers
Transitory	– Resource-poor households vulnerable to shocks especially, but not only drought – Pastoralists – Others vulnerable to economic shocks (e.g. in low potential areas)	– Urban poor vulnerable to economic shocks, especially those causing food price rises	– Groups affected by temporary civil unrest

Source: Development Studies Associates, 1998

The above classification has practical significance, i.e. it could guide the selection of types of intervention strategies. Households living under the spectre of chronic food insecurity (e.g. landless households in rural areas and low income households in urban areas) need different forms of interventions from those living under transitory food insecurity (e.g. farmers in drought-prone rural areas and urban poor vulnerable to economic shocks). A distinction has, for instance, been made by the then Ethiopian Disaster Prevention and Preparedness Commission between the strategies proposed against chronic (predictable) and acute (unpredictable) food insecurity: "… a long-term development and safety net is meant for chronic and emergency preparedness and response for the acute food insecurity" (DPPC, 2004:24). The next section presents the main causes of food insecurity (both chronic and transitory) in Ethiopia.

2.4.1. Causes of Food Insecurity

There are diverse views on the causes of food insecurity in Ethiopia. Some tend to focus on natural conditions such as bad climate or poor soil, while others see the main problems in socio-economic factors and government polices. Most scholars, nevertheless, identify a combination of many different factors as underlying reasons. Fassil G. Kiros (2005) proposes a framework of analysis of famine causation and the appropriate responses (Fig. 2.1). He argues that "no people would fall victim to famine unless they are predisposed to such an eventuality" (Fassil G. Kiros, 2005:65).

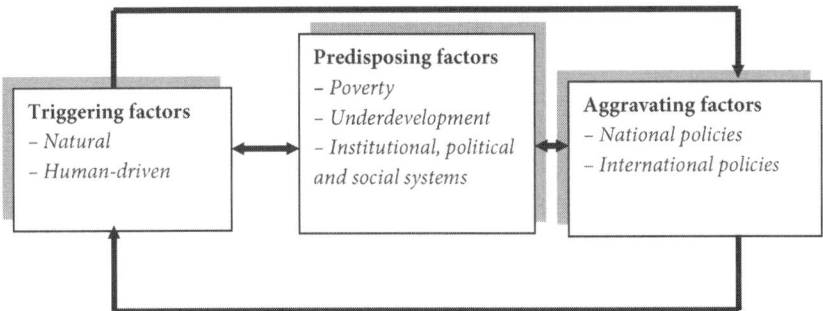

Fig. 2.1: Framework for analysis of famine causation and the appropriate responses (Source: Fassil G. Kiros, 2005:66)

The nature of the specific factors operating within each category can vary significantly depending on the political, economic, social and ecological conditions and circumstances characteristic of a particular country or region (Fassil G. Kiros, 2005). The succeeding paragraphs give a brief account of some of the factors that contribute to food insecurity and famine in Ethiopia.

Droughts
The Disaster Prevention and Preparedness Commission (DPPC) identifies drought-prone areas at national level using long-term meteorological data. Although almost the whole country is designated drought-prone, drought hazard has more negative impacts in northern and eastern Ethiopia, as well as in parts of the Rift Valley lowlands, than in other parts of the country. Many figures suggest that both the frequency and severity of droughts increased in Ethiopia in the last decades (Abebe, 2008; Haakansson, 2009). Parts of the country have, for instance, experienced droughts every year in the last decade, as opposed to every two to four years during the previous decades (Tassew Woldehanna, 2004).

Historical factors
As discussed in Chapter One, some scholars associate persistent poverty and food insecurity in Ethiopia with the system of landownership. This problem has prevailed during past regimes and is not yet fully resolved. During the imperial regime, land was mostly owned by absentee landlords. The land governance system thus "went far beyond a voluntary land-lease agreement and rather resemble[d] the extractive serfdom of the European Dark Ages" (Stellmacher, 2007b:524). Farmers had no incentive to use and manage land and other resources in a sustainable way, since land-use rights were frequently withdrawn without compensation.

Policies pursued by the subsequent regimes, namely the Dergue and EPRDF, are often argued to be essentially similar (land and the resources thereon are constituted to be under state ownership). It is well known that all rural and forest land was nationalised in 1975 by the socialist regime. The current government, led by the EPRDF, "also adopted a constitution in 1995 in which forests (land and other natural resources) [were] declared exclusive state property" (Gatzweiler, 2007). Some scholars argue that such "ideologically driven land policy" has stifled farmers' initiatives to adopt productivity-enhancing land improvement practices and induced the fragmentation of land holdings (IDCoF, 2002:xiii).

Inefficient and ineffective production systems and limited access to inputs
Agricultural production systems in Ethiopia have always been precarious. Even under 'normal' conditions, the per capita production in Ethiopia has never been much higher than subsistence level. This point was more extensively explored by Mesfin Wolde-Mariam (1984) in his famous book 'Vulnerability to Famine.' He makes the socio-economic and political system fully accountable for rural vulnerability to famine:

> A system in which the majority of peasants are totally dependent on the physical environment and on their backward methods of production, and in which the socio-economic and political forces persist in incapacitating the productive potential of peasants by incessant oppression and exploitation is a condition for vulnerability to famine (Mesfin Wolde-Mariam, 1984:169).

The low productivity can be attributed to a lack of agricultural technologies and inputs such as improved seeds, fertilizers, or pesticides, and a lack of access to formal credit and banking services which makes peasants dependent on prohibitive lending rates of private money lenders.

Taxation and an insufficient formal credit system

The multiple governmental taxes imposed on the peasants are also listed among the main causes of food insecurity and famine. Prior to the removal of the Dergue, all rural taxation, expressed as a proportion of gross agricultural income, amounted to 23 per cent for low-income households and 12 per cent for upper-income households (Getachew Diriba, 1995). Additionally, insufficient formal credit systems and bad storage and processing facilities force peasants to sell their products immediately after harvest. All peasants bringing their crops to the market at the same time causes drastic price fluctuations.

Instability and war

In the second half of the 20th century, Ethiopia was in a state of permanent conflict, internally and externally. The government allocated a remarkable proportion of the national budget to its military, and by 1990 the civil war consumed 60 per cent of the country's national funds (Horne and Frost, 1992). Such a chronic waging of war significantly contributed to food shortages and famine, because large amounts of manpower and resources that could have been used for rural development and improved household food security were shifted instead in other directions. The wars also claimed a tremendous amount of human life and property and devastated large areas of arable land, especially in the north of the country.

A study by de Waal (2000), focusing on the causes of food shortage and famine in northern Ethiopia during the 1983–85 food crises, appears to underline the role of war:

> Drought and harvest failure contributed to the famine but did not cause it. The economic and agricultural policies of the government also contributed, but were not central. The principal cause of the famine was the counter-insurgency campaign of the Ethiopian army and air force in Tigray and north Wollo during 1980–85. The zone of severe famine coincided with the war zone, and the phases of the developing famine corresponded with the major military actions (de Waal, 2000:115).

An international dimension: donor-driven policies

In developing countries like Ethiopia, there appears to be ready acceptance of new recommendations, especially those emanating from potential donors. Shifts from one type of development policy to another seem to be motivated largely by the amount of assistance obtained, rather than by what ultimate development outcomes can be realistically expected to result (Fassil G. Kiros, 2005). The same writer further contends that "[i]f there is one important lesson that developing countries should have learned by now, it is the fact that none of the international donors who usually come up with new development paradigms have a better knowledge of the needs and development potentials of all nations" (Fassil G.

Kiros, 2005:172). The proof of this, according to Fassil, is the very fact that they are prepared to abandon the policies they have advocated in the past in favour of new ones, even before the effects of the former have been evaluated.

2.4.2. Consequences of Food Insecurity and Famine

The consequences of famine in Ethiopia are manifold. Bahru Zewdie (1991:72) discusses some extreme consequences in the 19th century such as people eating the carcasses of animals, only to die painfully from the diseased meat, and some even resorting to cannibalism. The more recent problems of food insecurity and famine in the country have also resulted in a multitude of destructive impacts on the economy of the nation and well-being of her population. The following sections present a brief survey of the consequences of food shortages and recurrent famines, which have become part of the country's identity over the past decades.

Loss of means of production
Affected people tend to sell off their means of production to cope with the problem of food shortages and famine. Harrison (1990) reported that farmers in the central highlands of Ethiopia sold first their sheep and goats, then young cattle, mules and asses, then their cows, and finally their draught oxen. In some places even ploughs and hoes were reported to have been sold. A study in the Wollo region of northern Ethiopia revealed that a high percentage of peasants sold their dwellings at the high point of the crisis (Dessalegn Rahmato, 1988). Loss of means of production, in turn, reinforced the cycle of food insecurity and famine, as people failed to produce enough food when the conditions for production were reinstated.

Diverted development opportunities
The economic development of Ethiopia is affected by her inability to feed the population. Efforts and resources that could be used for sustainable agricultural development are too often diverted to cover the operational costs related to food aid distribution. Although the majority of these expenditures come from abroad (Tassew Woldehanna, 2004), the administrative and local transport costs are often covered by the Ethiopian state and non-state actors. For a period ranging from 1994 – 2003, the mean total expenditure on food aid in Ethiopia was 5.1 per cent of GDP and 17.4 per cent of total government expenditure, as shown in the following table.

Table 2.5: Total budget and food aid in Ethiopia

Year	GDP in million Birr	Total expenditure in million Birr	Expenditure on food aid as a percentage of GDP	Expenditure on food aid as a percentage of total expenditure
1994/95	33885	8372.2	6.81	27.54
1995/96	37937.6	9144.9	4.69	19.47
1996/97	41465.1	10016.9	2.16	8.94
1997/98	45034.9	11228	2.70	10.82
1998/99	48421	15148.3	3.93	12.58
1999/00	51869.4	17172.9	5.79	17.49
2000/01	51961.8	15864.6	10.52	34.45
2001/02	51158.1	17651.1	5.05	14.63
2002/03	57091.9	20715.9	4.0	10.99
Mean	**46,536.1**	**13,923.9**	**5.1**	**17.4**

Source: Tassew Woldehanna, 2004:40

Tragic loss of human life

Food insecurity and famine cause the loss of human life. The so-called 'Great Famine,' between 1888 and 1892, is estimated to have killed one-third of the country's entire population; in some areas to the north, the population was reduced by more than half (Mesfin Wolde-Mariam, 1984). A quarter of a million died during the 1973-74 famine (Girma Kebede, 1988), while the 1984/85 famine is estimated to have killed as many as one to one-and-a-half million people (The Economist Intelligence Unit, 1990; Mesfin Wolde-Mariam, 1991). In the recent past, the magnitude of death has reduced (due largely to increased production and continued food aid), although food shortage continues to affect millions of people in the country. It is hence argued that "[i]n place of the famine disasters of the eighties, hunger has become chronic; a natural part of everyday life for millions of Ethiopians" (Haakansson, 2009:14).

Socio-psychological repercussions

Besides the consequences described above, food insecurity and famine have social and psychological repercussions that are not that easy to quantify. These include "hopelessness that makes a human being empty to the core", "helplessness that tortures a human being standing face to face with slow but certain death", a "feeling of nothingness, being neglected and forgotten", and spatial dislocation and the accompanying dismemberment of families which, in turn, result in tearing and shattering of the social fabric (Mesfin Wolde-Mariam, 1984:55). Describing such costs of famine, Mesfin emphatically and unmistakably notes that "[i]t is impossible to estimate the money value of the daily helpless suffering of the mother whose pain of hunger is compounded by the innocently persistent demands of her children. It is impossible to estimate the money value of the tearless cries and the wrinkled faces of starving children. Day by day, misery is compounded by the debilitating process of starvation".

2.5. The Ethiopian Food Security Strategy

> There is little justification to invest in development activities which are only remotely connected to food and agriculture under conditions where the people are unable to benefit from them directly and immediately (Fassil G. Kiros, 2005:8).

Sustainable agricultural development, a cornerstone for ensuring food security, depends on capacity to maintain a proper balance between the conservation and utilisation of natural resources (IDCoF, 2002). The provision of alternative options for people to meet their basic needs is believed to be a "promising avenue to countervail the forces that work against sustainable management of natural resources" (IDCoF, 2002:x). With such an understanding, according to the same source, "past and current governments have initiated policies and strategies for the proper exploitation of natural resources". The agricultural development-led industrialisation (ADLI) policy of the current government focuses on improving the productivity of small farmers and pastoralists to combat food insecurity.

One should, nevertheless, note here that coordinated action to improve food security in Ethiopia started only in the latter part of the 1990s, when the EPRDF-led government developed a food security strategy and subsequent food security programme in 1996 and 1998, respectively. The strategy aimed at building the resource base of poor rural households, expanding employment and income opportunities and providing targeted transfers to households with special needs.

Building blocks of the national food security strategy
In general, the national food security strategy focuses on three aspects, namely:
- increasing food and agricultural production,
- improving food entitlement, and
- strengthening capability to manage food crises.

The food production component focuses on the diffusion of improved technologies in areas where there is ample rainfall. Major emphasis is put on improving productivity in subsistence smallholder agriculture. In the food entitlement strategy, the focus is on reducing vulnerability in drought-prone areas by carefully designing and executing poverty-reducing developments. The national food security strategy also focuses on strengthening emergency capabilities by maintaining emergency food security reserves, developing an effective early warning system and holding strategic seed reserves.

The New Coalition on Food Security

In 2004, the country's Disaster Prevention and Preparedness Commission (DPPC) substantiated that "there is a strong political will... to overcome the root causes of food insecurity within a 3 - 5 years period with the support of partners including donors, NGOs and UN systems" (DPPC, 2004:3). This plan, which was considered by some civil society organisations as overtly optimistic, proposed to use strategies such as intensive water harvesting, safety nets, alternative income generation activities and the resettlement of people from vulnerable to fertile locations, in order to reduce disaster risk in a sustainable manner. Social mobilisation was also considered as a key component of the so-called 'New Coalition on Food Security (NCFS).

The NCFS appears essentially to have the same objectives as the national food security strategy, which was launched in 1996. In the following, the main objectives of the NCFS are listed (DPPC, 2004:8).

– *Increasing availability of food by increasing food crops and livestock production/productivity through improving moisture conservation and utilisation; promotion of soil conservation, crop diversification/ intensification/specialisation; improving feed and water availability; and strengthening extension services for crop and livestock production.*
– *Increasing access to food by improving agriculture and non-agriculture incomes through enhancing a safety net programme; promotion of household income diversification; enhancing market effectiveness; improving knowledge, attitude and skill/practices and strengthening appropriate microfinance institutions.*
– *Promoting preventive and curative health care through improving preventive and curative health services.*
– *Providing access to land through a resettlement programme.*

A practical response: Safety net programmes

A safety net includes various transfer programmes aimed at both redistribution and risk reduction (Tassew Woldehanna, 2004). Such programmes provide support to individuals, households, communities or regions that would otherwise be forced to subsist at levels below the poverty line. Safety net programmes are also believed to prevent the poor from having to engage in criminal or marginalised activities, or from being forced to accept extreme insecurity and risk of death (Tassew Woldehanna, 2004). Publicly-provided safety nets also grant a way of ensuring that crises do not halt human development, so that household investments in health and education are maintained and the poor do not divest the physical and social capital upon which their present and future productivity depends.

Safety net systems can be formal or informal (Tassew Woldehanna, 2004). The former include state food subsidies, NGO feeding or employment programmes, social funds and child allowances. Informal safety nets include community-based arrangements and private safety nets that help mitigate deprivation and temporary income shortfalls. A system of labour transfers within communities, *Edir*[7], remittance and transfers among extended families are the most common forms of informal safety nets in Ethiopia.

2.6. The Case Study Areas

As indicated in Chapter One, four case studies have been selected from the EKHC's programmes/projects to analyse the chosen activities more closely. This section gives a brief background to the areas in which the projects operate.

2.6.1. Kembata Tembaro zone and Alaba Liyu Wereda

Two of the projects, namely school capacity building and alternative basic education, operate in the Kembata Tembaro zone and Alaba Liyu Wereda; their management offices are located in Awassa and Alaba towns, respectively. As can be seen in Figure 2.1, Alaba Wereda was part of the Kembata Alaba Tembaro zone. Recently, the Wereda has been given special district (*Liyu Wereda*) status, and hence is no loner part of Kembata Tembaro (KT). It is important to note here that most of the information in the next section is presented in an aggregate form, because the data was generated when the Kembata Tembaro zone and Alaba Liyu Wereda were part of the same administrative unit, i.e. the Kemabta Alaba Tembaro zone.

Natural resource base
The Kembata Tembaro zone and Alaba Liyu *Wereda* are located in SNNPR, and together cover an area of 2,383.28 km^2 and have a population of 835,815 (Fig. 2.1). About 94 per cent of the population live in rural areas and are mainly engaged in agricultural practices. While most of the Kembata Tembaro zone is found within the Omo–Gibe Basin, Alaba Liyu *Wereda* is located within the Great

---

7 *Edir* is a voluntary social organisation established by people living in a certain geographic locality to support each other during such problems as the death of family members (Stellmacher and Mollinga, 2009).

East African Rift valley system. Topographically, the Kembata Tembaro zone is characterised by steep slopes at the foot of the Ambericho, Dato and Ketta mountains and valley sides. The Alaba Liyu *Wereda* is, however, flat and lies mainly at the heart of the Ethiopian Rift Valley. Agro-climatically, 17 per cent of the total area of the Kembata Tembaro zone and Alaba Liyu *Wereda* is classified as *Dega* (humid sub-tropical), 70.4 per cent as *Woina Dega* (sub-tropical) and 12.6 per cent *Kolla* (tropical).

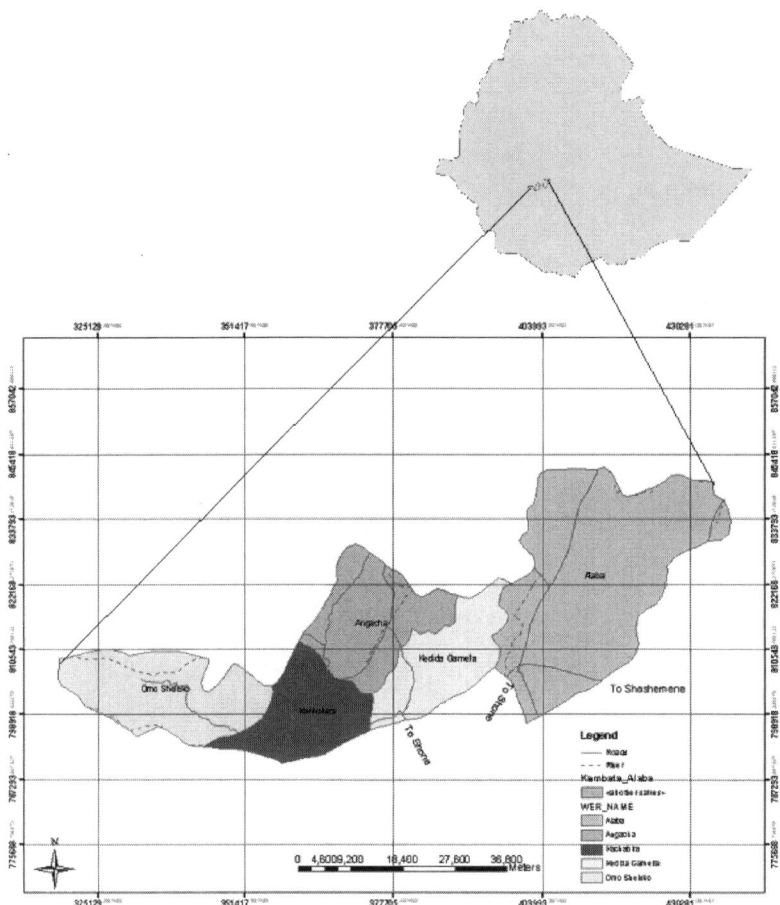

Fig. 2.2: Larger scale map of the study *Weredas*

The Kembata Tembaro zone is one of the most densely populated regions in the whole country, with an average density of about 310 persons per km^2 (DTRC, 1998). There is a severe shortage of arable land throughout the zone, which forces people to use marginal lands for cultivation. The average land-holding size is less than 1.5 ha (DTRC, 1998). Generally, agricultural productivity is very low, even compared to the other SNNPR zones. One of the consequences of this is the inability to attain food self-sufficiency. A study conducted by the Demographic Training and Research Centre at Addis Ababa University shows that in the five study zones of the SNNPR, 56.6 per cent of the households suffer from inadequate food reserves (DTRC, 1998). The percentage is found to be 95 per cent for the former Kemabta Alaba Tembaro zone (that includes Kembata Tembaro zone and Alaba Liyu *Wereda)*, which implies that only 5 per cent of the people in the project area produce sufficient food for the whole family throughout the year.

Fig. 2.3: *Enset* plantation in Angacha *Wereda*

Farmers in the region use crop rotation and inter-cropping, which have positive impacts on soil conservation. Furthermore, the planting of *enset* (false banana) reduces erosion through its large leaves and thick stems that restrain soil movement (Fig. 2.2).

The current status of the environment

Deforestation and soil erosion are the most serious and widely prevalent environmental problems in the Kembata Tembaro zone and Alaba Liyu *Wereda* (Fig. 2.3. and 2.4). There is no primary forest left in the areas except very small forests situated in Omo Sheleko and Kacha Bira *Weredas*. Acacias and eucalypt trees scattered in crop fields and on non-arable lands are the main sources of fuel. People also use cow dung and crop residues as supplementary sources of energy.

Fig. 2.4: House threatened by soil erosion

A study sponsored by the Economic Commission for Africa (ECA, 1997) identified a lack of biodiversity, over-utilisation of primary forest resources, shortage of land, farmers' reluctance in communal tree planting activities, shortage of trained manpower, logistical problems and a lack of replacement planting tradition as the main problems in the Kembata Tembaro zone and Alaba Liyu *Wereda*. More recently, the Rural and Agricultural Development Bureau of the respective *weredas* assessed the status of the environment (Aklilu Dalelo, 2006). This report shows soil erosion, deforestation, energy and water scarcity as the major environmental problems.

Fig. 2.5: Eucalypt plantation

2.6.2. Adaa Liben Wereda

As indicated in Chapter One, the Kuriftu children's care project operates in the Oromiya region, with its management office located in Kuriftu. The Kuriftu Centre can be found in Adaa Liben *Wereda*. The surface area of the Adaa Liben *Wereda* (now divided into two *weredas:* Adaa and Liben) is about 161,065 km^2 and the total population is 288,467 (in 2006). About 96 per cent of the population is classified as rural.

Babogaya (the area identified for implementing most of the activities related to family-based child sponsorship) is located about 65 km south-east of Addis Ababa and 13 km off the main Addis Ababa-Debre Zeit road. It is located at an altitude of about 1900 m above sea level and has a rainfall of 845 mm and mean, mean maximum and mean minimum temperatures of 16°C, 22°C and 10°C, respectively. The climate of the area is warm to cool *(Weinadega)*. Field crops such as wheat, chick peas, *guaya* and *teff* are the major sources of livelihood for the community, while the major religions of the area are Islam and Christianity.

Problems identified by the communities in the project area include environmental degradation, a high rate of unemployment, scarcity of arable land, soil erosion, water logging, food shortage, lack of feeder roads, etc. One-third of the children in the area are found to be malnourished. Women spend, on average,

12 – 14 hours a day preparing food, fetching water and firewood and working on the fields. This leaves very little time or energy for attending to the needs of their children. The baseline survey conducted before the implementation of the project also indicated that less than 30 per cent of school-age children attended school, because of a paucity of schools and the inability of parents to pay school-related expenses.

3. Faith-based Organisations and Development

3.1. Perspectives on Faith and Development

> If there is no engagement with the religion, any development work risks not only failing to address poverty, but also producing negative side effects of which the development worker may be tragically unaware (Hughes, 1998:148).

It appears that there is a growing interest in, understanding about, and recognition of the engagement of FBOs in development activities in less developed countries, such as Ethiopia. At the same time, scepticism persists on the underlying motives of FBOs to engage in these sectors at all. Some argue that they do so because it is an integral part of their mission. Others suspect that FBOs' development work is used as a strategy to urge people – especially the poor – strengthen or even change their faith. In the context of this work it is therefore important to explore the relationship between faith and development work in general, and the philosophical and theological basis for FBOs to participate in community development activities in particular.

Faith as a Social Capital
Social capital can be defined as features of social organisations, such as networks of individuals or households, and the associated norms and values that create externalities for the community as a whole (Putnam, 1993, quoted in Grootaert and Bastelaer, 2002:2). Empirical studies indicate that faith can be a form of social capital (Candland, 2001). Bornstein notes in a study conducted in Zimbabwe that faith is a force that motivated the employees of the FBOs Christian Care and World Vision "to do the work that they do and a force that brings them together" (Bornstein, 2002:26–27). It was further underlined that, more than mere rhetoric or ideology, faith provided a rubric "through which employees of the two NGOs interpreted the logic of development itself" (Bornstein, 2002:26–27).

Faith and Economic Progress
In 1968, Boulding presented a compelling argument on this issue in his book entitled "Beyond Economics". According Boulding, religious practices play a key role in shaping economic institutions, because "many other elements of the pattern of life, such as sex, child rearing, work habits, agricultural and industrial practices, are themselves profoundly affected by the prevailing religious beliefs" (Boulding, 1968:198). He adds that "for a very long time in human history, religion had a strong impact on not only the nature of the commodities which form the main subject matter of economic life, but also the economic institutions and practices which so largely determine the course of economic development" (Boulding, 1968:180). Boulding further highlights the impact of religion on the minds of people, thereby shaping their beliefs, customs and practices.

Religion as a Barrier to Development
Some scholars maintained, as thoroughly discussed by Geest (1993) and Marshall (2005), the view that religion is rather an obstacle to development and a barrier to modernisation and that religion and development can be seen as separate and incompatible realms.

Implications
The foregoing review shows that the relationship between faith and development is multifaceted over time and space. Some forms of faith are likely to encourage positive socio-economic transformation, while other forms may discourage development or even inject elements that spoil the fruits of development. The next sections briefly present the cross-section of religions vis-à-vis their potential to contribute to positive socio-economic transformation.

3.2. Factors Enhancing FBOs' Contributions

3.2.1. A Clear and Accommodative Mission

FBOs can make substantial and positive contributions to socio-economic development if their mission has clear and unequivocal provisions for such contributions. Reviews made on the mission of FBOs operating throughout the world indicate a wide assortment of interests ranging from purely religious to being more wholistic and accommodative (Braaksma, 1994; Clarke, 2005; Dicklitch and Rice, 2004; Geest, 1993). Clarke (2005:23) makes an important distinction between the four main ways in which FBOs deploy faith through social or political engagement, or link faith to development or humanitarian objectives.

- *Passive:* Faith is considered to be subsidiary[8] to broader humanitarian principles as a motivation for action and in mobilising staff and supporters and plays a secondary role in identifying, helping or working with beneficiaries and partners. Braaksma seems to strongly advocate this view: "[...] we did not see community development as just a means to Evangelism but as an essential component in the gospel message itself. Therefore, we felt that we needed to do such work with integrity which meant addressing the underlying causes of the problems the community was facing" (Braaksma, 1994:2).
- *Active:* In this case faith provides an important and explicit motivation for action and in mobilising staff and supporters. It plays a direct role in identifying, helping or working with beneficiaries and partners, although there might not be discrimination against non-believers.
- *Persuasive:* Here faith provides an important and explicit motivation for action and in mobilising staff and supporters. It plays a significant role in identifying, helping or working with beneficiaries and partners and provides the dominant basis for engagement. In this particular case, the desire to bring new converts to the faith, or to advance the faith at the expense[9] of others, is high and relatively more direct.
- *Exclusive:* According to this view, faith provides the principal or overriding motivation for action and in mobilising staff and supporters. It provides the principal or sole consideration in identifying beneficiaries. Social and political engagement are rooted in the faith and are often militant or violent and directed against one or more rival faiths.

Geest (1993) presents four models that explain the different approaches of FBOs to development. He underscores that these models are not distinguished on *whether* belief change or spiritual issues are part of the development process, but on *how* their proponents view the role of belief in development processes. The distinction between the models is therefore based on the way in which an organisation deals with the relationships between development and religious activities.

---

8 For instance, Dicklitch and Rice (2004:66) characterise the Mennonite Central Committee not to be a "traditional missionary organization. It does not seek to proselytize, although it works with mainly local faith-based organizations. In fact, it is the values that the MCC promotes and embodies rather than the promotion or practice of religion that makes it such a successful development organization"..

9 This model is also described as a colonial model: "[...] if a missionary is to do authentic community development work in areas which are primarily of another religion, it seems essential that such development be done in an inter-faith context. To do otherwise is to be perpetuating a colonial model, which is not true to the gospel, nor in the very practical sense is it productive" (Braaksma, 1994:54–55).

Model 1: Development without other religious activities.
Model 2: Development and other religious activities, with primacy assigned to development.
Model 3: Development and other religious activities, with each component clearly distinguishable.
Model 4: Development and other religious activities, with primacy assigned to other religious activities.

Table 3.1 shows the salient characteristics of the four models. One can notice a similarity between the four models of Clarke (2005) presented above and the four models by Geest (1993), which are described below.

Table 3.1: Models that explain approaches to development and other religious activities

Model 1: Development without other religious activities	Model 2: Development and other religious activity, with primacy assigned to development
– Development activities, whether accompanied by specific spiritual changes or not, are important on their own merits. – No activities are undertaken to convert people or support activities leading to conversion or church development. – More value is attached to "incarnational" or "presence ministry" than seeking to apply specifically Christian development approaches. – Evangelism is done indirectly through local churches in the project areas, but not by project staff/people supported by the project.	– Genuine development includes a change in spiritual circumstances as well as physical, social and political ones. – The desired outcome of spiritual change is acceptance of the Christian faith. – Evangelism and church planting are part of the development ministry. – Development is the primary purpose, and spiritual transformation is seen as a critical component of the development process. – All the activities underlying a wholistic development (e.g. health, education, agriculture) are given equal emphasis.
Model 3: Development and other religious activity, with each component clearly distinguishable	**Model 4: Development and other religious activity, with primacy assigned to other religious activity**
– An integrated approach combining distinct elements into one overall programme, but no intrinsic connection between development activities. – Development and Evangelism are distinct and separable tasks. One does not require the other for validity. – No one will convert to Christianity as a means to benefit from development activities. – There should be a distinction between mission or Evangelism, relief and development initiatives.	– The real mission of the church is to preach the gospel, with the aim of converting others to the Christian faith. – Development begins with converted people, i.e. once people abandon their traditional religions they are more likely to sustain a process of development. – Development work can be viewed as a pre-Evangelistic activity, the former used as a means to reach local people for the Christian faith.

In this context, Clarke (2005:12) defines five FBO categories, namely:
- faith-based representative organisations which rule on doctrinal matters, govern the faithful and represent them through engagement with the state and other actors,
- faith-based charitable or development organisations which mobilise the faithful in support of the poor and other social groups, and which fund or manage programmes that tackle poverty and social exclusion,
- faith-based socio-political organisations which interpret and deploy faith as a political construct, organising and mobilising social groups on the basis of faith identities but in pursuit of broader political objectives, or, alternatively, promote faith as a socio-cultural construct, as a means of uniting disparate social groups on the basis of faith-based cultural identities,
- faith-based missionary organisations which spread key faith messages beyond the faithful, by actively promoting the faith and seeking converts to it, or by supporting and engaging with other faith communities on the basis of key faith principles, and
- faith-based radical, illegal or terrorist organisations which promote fundamental or militant forms of faith identity, engage in illegal practices on the basis of beliefs or engage in armed struggle or violent acts justified on the grounds of faith.

In view of the contribution of FBOs to socio-economic development, one can thus hypothesise that FBOs that fall under the second category would have the highest chance of contributing to positive socio-economic transformation, while those that fall under the fifth category are likely to provide minimal contributions or even play counterproductive roles.

3.2.2. A Conducive International Atmosphere

Interest by policymakers and development scholars in FBO activities has increased over the last decades. In 1998 the World Faiths Development Dialogue (WFDD) was established by the then president of the World Bank and the Archbishop of Canterbury to help promote a dialogue on poverty and development, both among the different faith-based organizations, and between them and international development agencies (Marshall and Saanen, 2007; Tyndale, 2001:23). The WFDD was strongly backed by the Bush Jnr. administration. Concomitantly, the awareness of other governmental and non-governmental development concerned organizations in donor as well as in recipient countries on the role and potential of FBOs in development work increased (Belshaw, 2006).

In general, FBOs have gained an international reputation with regard to their ability to mobilise people and resources, and to be present in rural areas of less developed countries. The World Bank's research report, "Voices of the poor", substantiated that both urban and rural poor people value the guidance of religious and community-based organisations (www.pathfind.org). According to Marshall (2005:10) and Belshaw (2006:158), this can mainly be referred to through certain features and attributes of FBOs, listed below.
- *Profile:* The fact that in poor countries the majority of their membership is likely to consist of the poorest and most marginalized.
- *Philosophy:* Christianity and most other world faiths stress a variant of the 'golden rule' (treat others as you yourself wish to be treated) as a guide to social relationships.
- *Moral:* FBOs spur people to grapple with ethical issues ranging from corruption to equity.
- *Links and Networks:* The existence of separate intra-country and international links to sister-organisations possessing some funding and experience-providing capability. FBOs work through myriad peace-making channels, sustaining communities and spearheading the rebuilding and healing process.
- *Economy:* FBOs promote public support for development assistance and help forge consensus around hard choices.
- *Empowerment:* Spiritual and relational experiences can raise the self-regard and confidence of previously excluded poor people, helping them to benefit from new opportunities.
- *Sustainability:* FBOs have a long-term commitment to their members. They are likely to remain in place through a variety of difficult circumstances.

3.3. The Major Christian FBOs in Ethiopia

3.3.1. Background

Article 11 of the Constitution of the Federal Democratic Republic of Ethiopia, which came into force in 1995, states that "state and religion are separate", that "there shall be no state religion" and "the state shall not interfere in religious matters, and religion shall not interfere in state affairs"[10]. Article 27 deals with provisions related to 'freedom of religion, belief and opinion.' This Article also allows the establishment of institutions of religious education and administration in order to propagate and organise one's religion (see details in Box 4.1.)

---

10 The Constitution of the Federal Democratic Republic of Ethiopia. Addis Ababa, 1995.

The Major Christian FBOs in Ethiopia

Box 4.1: Ethiopian Constitution Article 27

> Freedom of Religion, Belief and Opinion
> *Everyone has the right to freedom of thought, conscience and religion. This right shall include the freedom to hold or to adopt a religion or belief of his choice, and the freedom, either individually or in community with others, and in public or private, to manifest his religion or belief in worship, observance, practice and teaching.*
> *Without prejudice to the provisions of sub-Article 2 of Article 90[11], believers may establish institutions of religious education and administration in order to propagate and organise their religion.*
> *No one shall be subject to coercion or other means, which would restrict or prevent his freedom to hold a belief of his choice.*
> *Parents and legal guardians have the right to bring up their children ensuring their religious and moral education in community with their own convictions.*
> *Freedom to express or manifest one's religion or belief may be subject only to such limitations as are prescribed by law and are necessary to protect public safety, peace, health, education, public morality or the fundamental rights and freedoms of others, and to ensure the independence of the state from religion.*

3.3.2. The Early Missionaries

Catholic and Protestant Christian missionaries that started educational programmes in other African countries did not do so in Ethiopia until the beginning of the 20th century (Hailu, 1975:146). The influence of the Ethiopian Coptic Orthodox Church made it difficult for European missionaries to work in Ethiopia. In 1904, Emperor Menelik in his attempt to modernize the country allowed Swedish and French missionaries to start educational activities in Addis Ababa (Hailu, 1975:147). Besides, a few Ethiopians were sent abroad under the auspices of missionaries to acquire 'western' education and language skills (Markakis, 1974:144).

Before the Italian invasion in 1936, European Catholic and Protestant FBOs tried to establish missionaries in Ethiopia, despite "the disdainful attitude of Ethiopians in general and the active opposition of the [Ethiopian Coptic Orthodox] Church" (Markakis, 1974:146). While proselytising was theoretically limited to non-Christian areas only, missionary educational activities were in-

---

11 Article 90 (sub-Article 2): Education shall be provided in a manner that is free from any religious influence, political partnership or cultural prejudices.

creasingly tolerated, and Emperor Haile Selassie's benevolent attitude encouraged the missionaries to devote their efforts to this field. Missionary sources proudly reported that "throughout southern and western Ethiopia various missionary societies opened a network of schools for the country people. For the first time in their history, they were being educated" (Atkins 2003:np). In addition to the mission schools and the schools run by the Ethiopian Coptic Orthodox Church, a number of private schools were established, often operated by foreign non faith-based organizations.

Table 3.2. shows the absolute numbers of non-state schools and their student population in Ethiopia for mission schools, private schools and Orthodox Church schools in the year 1968. In view of this study it is interesting to note the relatively high number of missionary schools particularly on the secondary level.

Table 3.2: Enrolment in non-state schools in Ethiopia in 1968

Level	Type					
	Mission		Private		Orthodox Church	
	Schools	Students	Schools	Students	Schools	Students
Primary[12]	238	44631	332	57648	117	17824
Secondary	19	1337	16	1643	3	336

Source: Markakis, 1974:149

Pankhurst (1966) observed that foreign missionary activities considerably expanded after WWI, with the result that missionary education became even more widely diffused than state education – in terms of geographic coverage not in number of schools.

3.3.3. Current Role of Evangelical Churches

Number and Distribution
Protestant missionaries have been active in Ethiopia since the late 18th century (Hughes, 1998:15). Today, all Evangelical churches in Ethiopia combined have 10.5 million members organised in 29,805 local administrative units. 43.4 per cent of the Ethiopian protestants are organized under the Ethiopian Kale Heywet Church; 39.4 per cent belong to the Evangelical Church Mekaneyesus (ECFE, 2005:39).

Geographically, Evangelical Christians are found mainly in the southern and western parts of the country. Oromiya and SNNPR have the highest absolute

---
12 Includes junior secondary level (seventh and eight grades).

numbers of Evangelicals in Ethiopia (see Table 3.3). The percentage of the total population that belongs to the Evangelicals is as high as 54 per cent in Gambella, 38 per cent in SNNP, 25 per cent in Benishangul Gumuz and 18 per cent in Oromiya regions. Since about three decades, Evangelical churches experience very high growth rates in terms of membership. The annual growth rate between 1993 and 2004, for example, was 4.6 per cent – much higher than the population growth rate (ECFE, 2005:45).

Table 3.3: Distribution of Evangelical churches by region

Region	No. of Population (2005)	No. of Evangelicals	Evangelicals as a percentage of the population	Region's share of Evangelicals
Tigray	4,113,000	4,176	0.10	0.04
Afar	1,330,000	5,223	0.39	0.05
Amhara	18,143,000	22,982	0.13	0.22
Oromiya	25,098,000	4,612,962	18.38	44.16
Somali	4,109,000	4,179	0.10	0.04
Benishangul Gumuz	594,000	148,333	24.97	1.42
SNNPR	14,085,000	5,325,469	37.81	50.98
Gambella	234,000	125,353	53.57	1.20
Harari	185,000	1,044	0.56	0.01
Addis Ababa	2,805,000	187,938	6.70	1.80
Dire Dawa	370,000	8,358	2.29	0.08
Ethiopia	71,066,000	10,446,017	14.7	100.00

Source: ECFE, 2005:41

Engagement in Development Works
As noted in previous sections, most faith-based organisations in Ethiopia consider social pro-poor services an integral part of their faiths; helping the poor is what basically all agree on. This appears to be vivid in the statement by Reverend Gudina Tumsa (Gudina Tumsa, 1975, quoted in GTF, 2003:67):

> The contribution the EECMY [Ethiopian Evangelical Church Mekaneyesus] makes to eliminate poverty, ignorance and disease is not to win favour from anyone. The EECMY does it because the need is there, and Christian love compels us to make as much contribution as possible, so that our country may be a beautiful and comfortable place to live in.

Although concern for the poor has always been present in the services of Evangelical churches in Ethiopia, the active involvement of Evangelical churches in long term community development projects/programmes is a relatively recent phenomenon. In fact, some Evangelical Christians used to consider development projects as 'sources of evil and immorality' and disapproved the implementation of such projects in their local churches or denominations (Aklilu Dalelo, 2003). The situation appears to have been changing since the early 1990s, though, as the two largest Evangelical churches, EKHC and EECMY, increasingly implemented development programmes that addressed a variety of issues ranging from environmental rehabilitation to advocacy against genital circumcision. In the following sections the development efforts of the EKHC are presented in more detail.

3.4. The Ethiopian Kale Heywet Church

3.4.1. Background

The EKHC was founded in 1927 by the Sudan Interior Mission (SIM, now renamed Serving in Mission). Since its foundation, the EKHC has grown to become the largest Evangelical church in Ethiopia, with more than six million members organised in about 7,000 local churches. Although one finds hundreds of EKHC churches dotted all over Ethiopia, EKHC has its geographic concentration in the southern and southwestern parts of the country,

Over the last 10 to 15 years, the EKHC Development Programme has grown in terms of both programme diversity and coverage. The focus has shifted, just as in the case of other FBOs, from relief and rehabilitation towards long term oriented sustainable development programmes (Aklilu Dalelo, 2003). EKHC now engage in activities ranging from spring capping to the establishment of modern irrigation schemes. However, this does not mean that the EKHC has entirely abandoned its relief operations.

3.4.2. Educational Services of the EKHC

Before the Dergue revolution in 1974, literacy programmes were carried out in almost every EKHC church in Ethiopia, mostly with both children and adults participating (Aklilu Dalelo, 2003). Despite often formidable resistance by landlords and government officials, these literacy programmes spread to elementary schools annexed to many of the local churches. The SIM ran the junior high schools, while the local EKHC churches took the responsibility of administering elementary schools (mainly up to grade 4). Henry Atkins, the then SIM Education Secretary, describes this system as follows (Atkins, 2003:np):

> When we were appointed to Wondo in 1959, the missionaries were confined to the mission station and the believers were being persecuted because there was little religious freedom and the Orthodox Church was opposed to missionary work. Blanche and I had been assigned to the elementary school, which had reached 4th grade. Despite the religious intolerance the Protestant churches grew rapidly around Wondo and the pressure for education increased. Our school was not large enough to meet the needs for more classes, so we agreed with the church [EKHC] that we would take 5th grade on the compound if they took 1st grade out in the communities. As they took 2nd grade, we would add 6th grade on the station until they had grades 1 – 4 and we had 5th–8th.

The efforts and contributions of the SIM missionaries to build the capacity of EKHC local churches in the southern and western parts of Ethiopia were remarkable. They assisted church schools by contributing part of the salaries for supervisors, provided books prepared in accordance with the specifications of the Ministry of Education and offered in-service training for teachers during the summer break. Flexibility was the key characteristic with respect to the structure and organisation of the church schools – some schools had a single (first) grade school, others had two or three or four grade schools. Some schools had benches, chairs and modern desks, etc., while others had only stones and logs for seats. In the beginning, classes were conducted inside the chapels, and separate houses were constructed gradually. What is more, the church developed a unique strategy to disseminate education to as many beneficiaries as possible. Those who completed lower primary education (grade four) were forced to give a teaching service for one year. This helped spreading the literacy programmes to remote rural areas of Ethiopia in relatively short periods of time.

The facts outlined above are in contrast to a study on NGO work in Wolaitta by Dessalegn Rahmato in which he writes that "prior to this period [the mid-1980s], the Catholic and Protestant churches, which had a presence here going back to the 1920s, had built one or two schools and established health services as part of their religious programmes, particularly in the latter part of the 1940s and early 1950s, but these investments were modest in terms of their impact on the population (Dessalegn Rahmato, 2007:45). According to a study by Aklilu Dalelo (2003), 120 primary schools where run by Protestant churches in Wolaitta with 182 teachers and 12,031 students when the Dergue took over all church schools in 1975.

Key challenges
Those serving in the EKHC educational services before 1974 still resent how landlords and government officials during the Emperors' time opposed the EKHC literacy programmes. EKHC was often accused of conducting their programmes without formal permission from the government. EKHC officials were forced to travel to Addis Ababa (partly on foot) to gain permission for their

initiatives (Mulatu Bafa, 2003). Similarly, SIM missionaries insisted that the EKHC only run elementary schools up to grade four. A comment by Mr. Eyoel Fugie, a former EKHC representative, substantiates this point:

> We expressed our desperate need to the missionaries to open secondary schools but they rejected the idea and limited local churches to run schools up to grade six. One of the missionaries undermined our request and bluntly told us to stop the dream to complete grade twelve in 'one generation.' Not giving up, we pressed our request for grade eight and above. Later on, they decided to open grade eight and handed over the responsibility of administering elementary schools to local churches (Aklilu Dalelo, 2003:29).

The above narrative is still shared by many in the EKHC. Whether it was the official policy of SIM or a stand taken by some missionaries requires further scrutiny. Following the 1974 revolution, all the schools, which were run by the local churches and SIM, were taken over by the Ministry of Education. The government confiscated the property of the churches and chased the church leaders and missionaries away, leaving them no chance of even playing an advisory role in school administration. Many schools were closed down. This hampered educational development in Ethiopia in general, and in the southern and southwestern parts in particular.

3.4.3. Current Situation and Future Direction

In 2006, the EKHC conducted a study to assess the overall situation of the church, before designing its second strategic plan (EKHC, 2007). The study shows the different development related activities and achievements over the years. It also identified topics that need more emphasis in the future. The main findings of the study are listed in the following.
- 5.7 million people were EKHC members in September 2006. They were organized in 6000 local churches.
- The total number of EKHC members increased by 25.5 per cent between 1991 and 2006.
- 27.2 per cent of the EKHC churches were engaged in literacy programmes.
- 42.5 per cent of the EKHC churches run some form of income-generating programmes. These programmes were, however, often not very effective due to lack of management capacities.
- Youths actively participated in leadership roles, while women's participation in leadership was low.
- The integration and transparency among the different organizational levels of the EKHC and its different departments was low. Information sharing about

externally funded projects was, for instance, limited within specific departments at headquarters.

Based on the results of this survey and comments gathered during planning meetings a EKHC strategic plan was designed for 2007 to 2012. 'Capacity building' (including educational development) and 'improving food security' were identified as two of the six key focal areas. The other areas were Evangelism and discipleship; children, youth and women development; health and HIV and communication and networking (EKHC, 2007). In the following, the education and development related activities of the 2007–2012 strategic plan are listed.
- Offering short- and long-term training to 320,000 people on issues related to leadership and wholistic development.
- Establishing 83 training centres, 95 primary schools and one teacher training institute.
- Planting one million fruit and other trees.
- Establishing 10 grain mills to reduce women's drudgery.
- Drilling 270 boreholes to supply drinking water.
- Installing 16,000 bio-sand filters to purify dirty water.
- Irrigating 680 ha of land by diverting small rivers.
- Providing training on health and sanitation to 123,102 people and constructing 20,000 latrines.
- Conducting campaign and awareness raising workshops on harmful traditional practices, and establishing anti-FGM clubs. This was planned to prevent FGM from being practiced on half a million girls.
- Starting and/or strengthening literacy programmes in 3,000 local churches;
- Proving basic educational and medical support to 15,000 orphans and 20,000 destitute children.
- Creating job opportunities for 300,000 women and youths by engaging them in local church-based literacy programmes and income-generating schemes;
- Mainstreaming HIV/AIDS prevention and control in 200 junior and 45 higher level bible schools.
- Organising 31,600 beneficiaries in self-help groups and carrying out capacity-building programmes.
- Constructing 2,000 low-cost houses for homeless families.

The above list shows the wide range of activities of the EKHC projects/programmes. One can see that concerns about educational development and food security are duly addressed, both directly and indirectly. Almost all the activities presented above are directly or indirectly related to improving household food security in the project areas. How far these ambitious plans will be practically realised is not yet clear and should be subject to future evaluation.

4. Faith, Ecological Balance and Food Security: Survey Findings

4.1. Interviewees' Profile

Besides the analysis of secondary literature and internal EKHC documents, this study empirically assessed the views and knowledge of EKHC representatives on the contribution of EKHC projects/programmes on educational development and ecological balance in Ethiopia. As indicated in Chapter One, a total of 206 people were interviewed by using a structured questionnaire (see Appendix I).

Table 4.1: Participants' main responsibility in the EKHC (area of service)

Area of service	No.	%
Evangelist	39	18.9
Spiritual program coordinator	29	14.1
District church leader	24	11.7
Distinct church secretary	23	11.2
Local church elder	20	9.7
Development worker	10	4.9
Member	10	4.9
Trainer/facilitator	8	3.9
Pastor	8	3.9
Bible school teacher	7	3.4
Youth ministry coordinator/singer	7	3.4
Women ministry leader	5	2.4
Finance staff	2	1.0
No response	14	6.8
Total	206	100.0

Table 4.1 shows that the study aimed to include interviewees with diverse responsibilities within EKHC. Full-time workers/'ministers' (Evangelists, pastors, Bible school teachers) account for 26.2 per cent of the interviewees. District-level chairpersons, secretaries and spiritual programme heads are the three most important positions in the church's administrative structure. They account for

37 per cent of the interviewees. These two groups make up the most influential strata as far as decision making in the church is concerned. Only 13 (6.3 per cent) of the interviewees are female. This very small number reflects the low representation of females in leadership positions in Ethiopian FBOs in general, and the Ethiopian Kale Heywet Church in particular.

Table 4.2: Participants' level of education

Level	No.	%
Grade 12 or below	23	11.2
Certificate (12+1)	26	12.6
Diploma (12+2)	98	47.6
BA/BSc	58	28.1
MA/MSc or above	1	0.5
Total	206	100.0

76.2 per cent of the interviewees hold an undergraduate academic degree. Nearly half (47.6 per cent) are diploma holders, while more than one-quarter (28.1 per cent) hold a BA/BSc, mostly in theology. Only around one-tenth of the interviewees fell under category 'grade 12 or below.' Although there are people who hold second and terminal degrees (mainly in the headquarters), the educational profile of EKHC representatives appears not to be commensurate with the size of the Church and the number and complexity of programmes undertaken.

4.2. Views on the Human Impact on the Environment

> Being knowledgeable in the field of science is nothing but understanding the laws of Nature. A judicious understanding of the laws of Nature and living according to their dictate is as rewarding as knowing the Word of God and living by it. Our clergy should accordingly possess awareness of science and philosophy as well as material development and progress (Sirgiw Gelaw, 2007:32).

The participants were asked to assess a statement strongly emphasising the value of the natural environment: "Nature must be saved or we humans will die". Nearly one-quarter of the respondents strongly agreed, while more than half (57 per cent) expressed their agreement (either agree or strongly agree). This shows the value that the interviewees attach to the natural environment. With regard to the negative impact of development on the natural environment, most did not agree to a 'development is destruction' view. More than half of the respondents (58.2 per cent) rejected the statement "Development activities are causing massive destruction to the natural system" (Table 4.3). Interestingly the interviewees do not value the role of science and technology in improving the current state of the natural environment particularly high. 39.7 per cent reject the statement "Science and technology can change the current state of degradation of the natural environment". This is a high figure given the fact that 76 per cent of the interviewees hold an academic degree.

Table 4.3: Participants' views about man's impact on nature

S/n	Statements	Respondents' views (percentage)						
		SA[13]	A	U	DA	SDA	SA+A	SDA+DA
1	How do you assess the statement: "Nature must be saved or we humans will die"?	24.5	32.5	16.0	16.5	10.5	57.0	27.0
2	Development activities are causing massive destruction to the natural system.	5.4	17.8	17.8	32.7	26.2	23.2	58.2
3	Science and technology can change the current state of degradation of the natural environment.	14.2	25.0	21.1	25.5	14.2	39.2	39.7

4.3. Views on the Link between Environmental Protection and Food Security

The findings indicate a high level of understanding about the relationship between environmental protection and food security. More than two-thirds (70.8 per cent) of the respondents could present a meaningful link between the two factors (Table 4.4). A closer look into the responses shows that more than 40 per cent of the participants had 'comprehensive[14]' or 'very comprehensive' explanations for the relationship. Some respondents tried to express their views in a figurative manner. The analogies 'hand in glove' and 'two sides of a coin" were frequently used to describe the relationship between environmental protection and food security.

---

13 SA refers to "strongly agree", A to "agree", U to "undecided", DA to "disagree" and SDA to "strongly disagree",
14 Comprehensive responses here refer to those statements that show the complex relationship between environmental protection and food security using more than one factor.

Table 4.4: Depth of analysis on the relationship between environmental protection and food security

Response category	No.	%
Not clear[15]	54	26.2
Not comprehensive	62	30.1
Comprehensive	59	28.6
Very Comprehensive	25	12.1
No response	6	2.9
Total	206	100.0

The following are examples of responses under the four categories: 'not comprehensive,' 'comprehensive,' 'very comprehensive,' and 'figurative.'
Not Comprehensive:
- *"Food is a result of protection of nature and care for the environment" (Evangelist from Kembata)*
- *"There is no food security without protection of the environment (Evangelist from Hadiya)*
- *"... environmental protection is a precondition for ensuring food security" (Evangelist from Kembata and Hadiya)*
- *"Environmental protection is the foundation of food security, while food security can be considered an outcome of protection and proper use of the environment" (Church staff/ 'minister' and school teacher from Jimma)*
- *"Environmental protection creates a favourable condition for development works aimed at ensuring food security" (Church elder from Jimma)*

Comprehensive:
- *"Proper management of the environment helps to reduce misuse and pollution of natural resources, and thereby ensures food security" (Evangelist from Aleta Wondo)*
- *"Protecting the environment improves climatic conditions, and improved climate helps the farmer to grow crops on time and this, in turn, ensures food security" (Evangelists from Dilla and Shashogo)*
- *"Natural resources are the basis for ensuring food security and fulfilling the other needs of human beings directly or indirectly. Any activity to protect natural resources will therefore have an impact on food security" (Youth service/ 'ministry' coordinator from Addis Ababa)*
- *"Food security depends on a protected and balanced environment. So they are*

---

15 Many respondents tried to define environmental protection and food security (in most cases correctly), but failed to show the link between the two.

linked directly. Adversity in one brings about a danger on the other" (Development project coordinator, Gamo)

Very comprehensive:
- "The environment is a pool from which we fetch everything. If the natural environment is not protected properly, our food security will be in danger. Food production and ensuring food security are possible only if the environment is protected. The two have a direct relation" (Church staff/ 'minister' and school teacher from Yanassie Afrara)
- "Food is what we get from the environment, the protection of which has been mandated by God to mankind. It is therefore essential to develop, protect and conserve our environment. An environment subjected to a process of degradation would react harshly in a way currently witnessed in many parts of the world" (Theologian and spiritual programmes coordinator from Ofa District)
- "Protecting the environment reduces the danger of land degradation and improves land productivity. Improved land productivity increases crop production. Increase in crop production improves food security. On the other hand, protection of water sources and supply of clean water reduces the prevalence of waterborne diseases. This, in turn, reduces expenses for medication, which are often covered by selling agricultural produce. Not selling off crops for covering medical expenses improves food security" (Theologian from Kucha[16])

Figurative:
- "The two can represent the two sides of a coin. Protecting the environment increases productivity of the environment. This will, in turn, increase the yield from nature" (Church staff/ 'minister' and school teacher from Kembata and Hadiya)
- "A protected environment is a foundation for food security, and expecting food security in unprotected environment is just like expecting rain from a cloudless sky" (Evangelist from Soddo Zuria District)

4.4. Views on Causes of Food Insecurity

As indicated in Chapter Two, many parts of Ethiopia are in a state of chronic food insecurity. In this survey, an attempt has therefore been made to assess the extent to, and ways in which, leaders of the EKHC and those working in EKHC's development projects perceive the causes of food insecurity. The participants of the study could identify about twenty factors as causing food insecurity in

---

[16] It is important to note here that a project dealing with environmental protection and food security has been implemented in the Kucha area by EKHC for more than ten years.

Ethiopia. The top seven factors, which were mentioned by more than 10 per cent of the respondents, are listed in Table 4.5. Undesirable or a bad work ethic (laziness) were cited as major reasons by nearly half of the participants, most of whom thus appeared to put the blame not on internal politics or external forces, but on members of their communities. Contrary to this viewpoint, academic debates on the causes of food insecurity and famine in Ethiopia or elsewhere tend to point fingers at national political inaction and/or international economic injustices (Aklilu Dalelo, 2001).

Table 4.5: Causes of food insecurity

s/n	Causes	No.	%
1	Undesirable work ethic (laziness)	98	47.6
2	Poor knowledge and production techniques	86	41.7
3	Natural hazards like drought, flood, untimely rains, etc.	62	30.1
4	Shortage and/or human-induced degradation of natural resource base (deforestation, overgrazing, etc.)	62	30.1
5	Unwise use of resources (wastefulness), poor planning	36	17.5
6	Lack of good governance/corruption	30	14.6
7	High dependency: too many consumers but too few producers	27	13.1

The second most important cause of food insecurity, according to the respondents, is poor knowledge and production techniques. Smallholder farmers in Ethiopia often still use almost archaic means of agricultural production like the hoe, and the rate of adult illiteracy is one of the highest in the world. About one-third of the respondents consider natural hazards like drought, untimely rains and flood contributing factors to the problem of food insecurity, while a further third mentioned increasing scarcity of resources and environmental degradation, induced primarily by human beings (like deforestation, overgrazing, etc.).

Close to one-fifth of the respondents believe that unwise use of resources (agricultural products or money) and poor planning are the other factors exacerbating the problem of food shortages in Ethiopia. Lack of good governance/corruption (mentioned by 14.6 per cent) also plays its part in keeping the poor food-insecure. The last factor mentioned by 13.1 per cent of the respondents is the high rate of dependency, i. e. too many people who are "able to eat but not able to work/produce", according to one respondent. Given the high percentage of population below the working age, this statement appears to make a lot of sense in Ethiopia.

The following are the other factors mentioned by less than 10 per cent of the respondents. The figures in brackets show the number of respondents who indicated the specific factor.

Views on Causes of Food Insecurity   75

- *Political instability and frequent changes in policies (16)*
- *Sense of dependency on external assistance (11)*
- *Heavy dependence on rain for agriculture (11)*
- *Low place/status given to farming (agriculture considered an illiterate person's profession) (7)*
- *Conflicts, internal and external (6)*
- *Exporting food while people die of hunger (5)*
- *Very selective and conservative food culture (4)*
- *Joblessness (4)*
- *Small size (fragmentation o)f farmland (3)*
- *Disobedience to the will of the Creator (3)*
- *Lack of capital to initiate small businesses (3)*
- *Lack of market for products (exploitation by middle-men/merchants) (2)*
- *Shortage of chemical fertilizers (1)*
- *Lack of land tenure/ownership right (1)*

Participants were also asked to express their views on the following statement: "Food insecurity in Ethiopia is more a problem of a lack of enough food within the country than a lack of access to food" (Table 4.6). This statement, which reflects the views of scholars working on food insecurity in Ethiopia (see Chapter Two), was rejected by half of the respondents (51.9 per cent), showing that at least that proportion of the respondents do not think that 'lack of access' is the major culprit.

Table 4.6: Participants' views about aspects of food security

S/n	Statements	Respondents' views (percentage)						
		SA	A	U	DA	SDA	SA+A	SDA+DA
1	Food insecurity in Ethiopia is more a problem of a lack of enough food within the country than a lack of access to food.	11.8	23.0	13.2	32.8	19.1	34.8	51.9
2	In Ethiopia, FBOs have a key role to play in building the capacity of communities to attain food security.	48.5	42.1	4.0	5.4	0.0	90.6	5.4

4.5. Views on Strenghts and Special Contributions of FBOs

The vast majority of the respondents (90.6 per cent) expressed their agreement with the statement: "In Ethiopia, FBOs have a key role to play in building the capacity of communities to attain food security" (Table 4.6). When it comes to the special roles played by FBOs, the agreement went down to half (Table 4.7). Areas where FBOs have special roles include, according to the respondents, training and educating community members on issues like the use of local resources and ways of becoming self-supportive. The focus on training related to 'hard work and use of local resources' seems to reflect the belief on the part of the respondents that the major cause of food insecurity is unfavourable work ethics. Nevertheless, only 12.1 per cent of the respondents indicated that FBOs have succeeded in bringing about positive changes in people's attitudes towards work.

Table 4.7: Strengths and special contributions of FBOs

s/n	Special contributions	No.	%
1	Training and educating communities to use local resources and become self supportive	104	50.5
2	Doing relief and development work along with the spread of the gospel	68	33.0
3	Mobilising members to live out their roles as responsible citizens and become examples of high moral standards	50	24.3
4	Proximity to the grassroots and ability to reach people in remote areas	36	17.5
5	Providing education and health services	46	15.5
6	Prayer and spiritual support	30	14.6
6	Bringing about changes in attitudes towards work	28	12.1
7	Advocacy to control harmful traditional practices and HIV/AIDS	23	11.1
8	Ability to forge international networks and raise funds	17	8.25
9	Strong social bondage, relatively better trust among members	10	4.85
10	Supplying clean water	4	1.94
11	Efforts being made to introduce development education in Bible schools	4	1.94

One-third of the respondents mentioned relief and development work as a strength of FBOs, whereas mobilising members to live out their roles as responsible citizens and become examples of high moral standards is mentioned by about one-quarter. Conversely, only 15 per cent of the respondents thought of prayer and spiritual support as a special role of FBOs. Advocacy to control harmful traditional practices and HIV/AIDS was mentioned by 11.1 per cent of the respondents.

Interestingly, some respondents stated that nothing had been done by FBOs to improve socio-economic conditions of their respective communities. A church leader from Soddo, who is an accountant by profession, said that "given the responsibility entrusted to them, it is difficult to say that they [FBOs] are doing anything special to bring about change". A nurse from Jimma, who was also serving as a deacon, stated that "So far, I have seen no strengths; all of what I see is FBOs quarrelling with each other". A similar view was shared by an Evangelist from Hadiya and the spiritual programme coordinator for Gamo District.

4.6. Views on Weaknesses of FBOs

The study also asked church leaders and development workers about the major weaknesses of FBO' work in Ethiopia (see methodological details in Chapter One). The most frequently cited weakness was the 'tendency to spiritualise everything, thereby undermining physical and social services.' Related to this problem is the tendency of some FBOs 'to make false dichotomy between the spiritual and worldly,' thereby making no effort to influence the 'other world.' Half of the participants of this study stated that such an attitude prevails among Ethiopian FBOs. 'Lack of proper communication between different denominations' is another bottleneck hindering efficiency and effectiveness in development endeavours undertaken by FBOs in Ethiopia (Table 4.8).

Table 4.8: Major weaknesses of FBOs

s/n	Major weaknesses	No.	%
1	Tendency to spiritualise everything, thereby undermining physical and social services	82	39.8
2	Lack of proper communication between different denominations	45	21.8
3	Tendency to leave social and development work to government or secular institutions	29	14.1
4	Encouraging dependency on donors	28	13.6
5	Lack of integrity, unfaithfulness	22	10.7
6	Making little effort to influence the 'other world'	22	10.7

Some respondents criticised FBOs to do 'more talk than action.' Others said that social/development work is not their business – an attitude whereby they 'let the government worry about it'. The coordinator of the Women's ministry at the EKHC National Office contended, for instance, that "there are still some EKHC churches which consider involvement in social and economic activities as a sin". A member of the Executive Committee of the Central zone critically stated that "Except a few, FBOs in Ethiopia are not in a position to change anything. In fact, one can say that their development efforts are bound to fail before reaching anywhere".

FBOs in Ethiopia are increasingly shifting their attention away from grass-root-initiated, rather isolated and small scale actions to donor-driven, more comprehensive and larger scale development work. Some respondents observe an increasing lack of integrity. One commented, for instance, that development work is making "only a few individuals wealthier and fatter".[17] For an Evangelist from Durame Town, the emerging problem is the lack of servant leaders and the increasing tendency to be "engulfed by a desire for modernisation and vying to quench such a desire". According to another Evangelist from Humbo District, the root cause of most of the problems of FBO' development work is "putting up leaders who have no vision and ability to lead their communities towards desirable social and economic change".

Other criticism expressed by the interviewees towards development-oriented FBO work is listed below. The figures in brackets show the number of respondents who gave the specific response.
- *Shortage of and/or inability to mobilise Christian professionals with the required vision and motivation (14)*
- *Lack of ability to manage relief and development (6)*
- *Wrong teachings: 'prosperity gospel' (5)*
- *Too many spiritual programmes, poor time management (5)*
- *Concentration in only a few localities, mainly urban areas (5)*
- *Less regard given to sustainability (4)*
- *Considering collaboration with government as an act of party politics (3)*
- *Increasing sense of conflict among members; declining love (2)*
- *Giving more attention to development work than spiritual services (2)*

4.7. Views on Evangelicals and Work Ethics

The renaissance of FBO' development work after the downfall of the Dergue in 1991 promoted a sustained discourse on the role of Evangelical churches in inculcating favourable work ethics. Some people argued that Evangelicals, in general, have less reputation when it comes to work ethics. Other allegations, sometimes by government officials, say that members of Evangelical churches (particularly youths) spend too much time chasing after workshops and conferences organised by FBOs in Ethiopia. As the issue has direct relevance to this study, the views of church leaders and people working in FBOs' development projects/programmes are shown in Table 4.9.

---

17 One of the reviewers of this study commented that "many of church members are becoming politicians just to make better salaries"

Table 4.9: Views about work ethics and capital accumulation

S/n	Statements	Respondents' views (in %)						
		SA	A	U	DA	SDA	SA+A	SDA+DA
1	Evangelical churches in Ethiopia strongly preach about hard work, but do not practice it.	26.3	42.9	14.1	15.1	1.5	69.2	16.6
2	Evangelicals in Ethiopia have a better reputation regarding hard work compared to followers of the Ethiopian Orthodox Church.	16.6	31.2	23.4	21.0	7.8	47.8	28.8
3	The Evangelicals in Ethiopia have a better reputation regarding hard work compared to the followers of Islam.	10.7	21.0	18.5	37.6	12.2	31.7	49.8
4	Evangelicals in Ethiopia tend to discourage business and investment.	5.4	20.6	20.6	31.4	22.1	26.0	53.5
5	Evangelicals in Ethiopia tend to discourage the accumulation of money and capital goods.	8.9	25.7	24.3	27.2	13.9	34.6	41.1

Table 4.9 shows that the great majority (69.2 per cent) of the respondents share the view that "Evangelical churches in Ethiopia strongly preach about hard work, but do not practice it". This finding is all the more interesting as all interviewees are themselves members of the EKHC. One the other hand, nearly half of the respondents (47.8 per cent) believe that "Evangelicals in Ethiopia have a better reputation regarding hard work compared to followers of the Ethiopian Orthodox Church". With regard to the comparison between Evangelicals and Muslims only 31.7 per cent agreed to the statement "Evangelicals in Ethiopia have a better reputation regarding hard work compared to the followers of

Islam". More respondents disagreed than agreed with the views that Evangelicals in Ethiopia tend to discourage business, investment, the accumulation of money and capital goods (Table 4.9).

4.8. Concluding Remarks

An attempt has been made to summarise the results of the survey with regard to the extent to, and ways in which, EKHC leaders and staff working in EKHC development projects view the relationship between environmental protection and food security, as well as the development-oriented work of FBOs.

First, most of the respondents seemed to value both natural environment and development. A contradiction between economy and ecology, often discussed in contemporary sustainable development debates, is not mirrowed in their views. One can see such a tendency from two angles. In its positive sense, the respondents do not think that development inevitably results in the depletion and destruction of the natural environment. In its negative sense, the respondents do not see the negative side effects of economic development. This might be due to their desire to fight poverty at any cost.

Second, there is a fairly high level of understanding about the relationship between environmental protection and food security. More than two-thirds of the participants understood that protection of the environment is one of the prerequisites of ensuring food security. Given the role FBOs play in mobilising their communities to engage in socially relevant activities, such a high level of understanding could enhance any action that may be taken to address Ethiopia's problems.

Third, participants of the study were able to identify diverse factors that cause food insecurity. However, their views appear to be skewed towards socio-cultural factors. More emphasis is, for instance, placed on such factors as unfavourable work ethics and use of backward techniques of production. Issues like political instability and fragmentation of farmland have been indicated as causes of food insecurity by a small number of participants. One can thus see a noticeable difference in the views of the participants of this study and those of scholars, who make government policies more accountable for food shortages in Ethiopia (see Chapter Two).

Fourth, the participants seemed to have mixed views on the development-oriented work of FBOs. The respondents believed that FBOs' contributions in the areas of training and provision of emergency relief are particularly helpful. However, they listed more weaknesses than strengths related to the services of FBOs. The weaknesses ranged from FBOs' tendencies to spiritualise every move they make to declining integrity – a practice that collides against what they stand for. In general, the findings of this study show that the members and staff of the EKHC are less optimistic regarding the potential of the EKHC to contribute to sustainable development in Ethiopia. This is particularly relevant given the very ambitious goals of the EKHC strategic plan 2007 – 2012 as shown in Chapter Three.

5. Case Study One: Empowering Schools to Address Ecological Balance

5.1. Background

Environmental education is one instrument for the transition towards more ecological sustainability (Palmer, 1998). Kofi Annan, the former UN Secretary General, made a remarkable comment with regards to the contribution of education to environmental protection: "Only by understanding the environment and how it works can we make the necessary decisions to protect it. Only by valuing all our precious natural and human resources can we hope to build a sustainable future".[18] This holds particularly true for children. Accordingly, the role schools and teachers play in environmental protection and natural resource management is receiving due emphasis (WCED, 1987; UNESCO and the Government of Greece, 1997; Fien, 2006).

Teachers and school administrators can play a major role in bringing the struggle for sustainable development into local communities around the world (UNESCO and the Government of Greece, 1997). Teachers can be considered key agents "for bringing about the changes in lifestyles and systems we need" (Fien, 2006:124). Similarly, the world's teachers are believed to have "a crucial role to play in bringing the message of the World Commission on Environment and Development (WCED) to the youth" (WCED, 1987:xiv). Teachers, however, can play these roles only effectively if they have, amongst other things, a curriculum which integrates sustainability and environmental issues, as well as the concerned knowledge and skills and willingness.

Several studies conducted on environmental education in Ethiopia indicate mixed results (Melaku Girma, 1994; Aklilu Dalelo, 1998, 2001, 2006; Abinet Gebrekidan, 2006; Shumete Gizaw et al., 2006). Aklilu (2001), for example, carried out a study on students' awareness of, and their attitudes towards the problem of natural resource degradation. More than 1,100 students took part in the study from junior and senior secondary schools of the then Kembata-Alaba-

---

18 Source: http://news.bbc.co.uk/go/pr/fr/-/1/hi/sci/tech/4391835.stm

Tembaro zone and the Awassa Teacher Training Institute, both in southern Ethiopia. The study revealed considerable deficiencies in students' awareness on issues related to natural resource conservation, use and management. Only one-quarter of the students were, for instance, able to give three of the causes, consequences and solutions to the most serious environmental problem in Ethiopia, i.e., land degradation. The study thus recommended the provision of a basic and aim-oriented education on issues pertaining to the use and management of natural resources in schools and teacher training institutions, so as to save the environment and improve the livelihood of its inhabitants.

In an attempt to put this recommendation into practice, a pilot project on school capacity building was initiated and implemented in SNNPR by the Capacity Building and Community Empowerment Programme of the EKHC. Its management office was located in Awassa town (Fig. 2.1). It run from 2004 to 2007 with financial and technical support from Serving in Mission (SIM) Canada and the Canadian International Development Agency (CIDA).

The target groups and beneficiaries of the pilot project were primary school teachers and students in the Kembata Tembaro zone and Alaba Liyu *Wereda*. The following were the major activities planned to achieve the aims of the pilot project.
- Undertaking a baseline survey on the status of the environment and the place of environmental issues in the schools' curriculum.
- Preparing a handbook on natural resource management and environmental protection.
- Conducting training for primary school teachers on the use and management of natural resources and environmental protection.
- Integrating issues related to the use and management of natural resources and environmental protection into the daily lessons.
- Establishing environmental clubs in all the project schools.
- Establishing centres for the development and dissemination of alternative energy technologies.
- Establishing school nurseries.

5.2. The Project Schools

The identification of the project schools was made based, among others, on the survey of their environment (see Chapter Two). A total of 11 schools (10.6 per cent) were selected by *Wereda* Capacity Building, Education and Rural Development offices (Table 5.1). The criteria used for the selection include proximity to water sources, the availability of space for planting, accessibility, capacity of school administration, proximity to other schools and the magnitude and seriousness of environmental problems. The project team went to each of the selected sites and checked whether the chosen schools really met the criteria used. This was achieved through observation and interviews with the community, school directors and teachers.

Table 5.1: Distribution of primary schools by *Wereda*, 2005

Wereda	Number of upper primary schools	Schools selected for intervention
Alaba	18	Gerema, Besheno and Mekala
Angacha	30	Kerekicho and Adancho
Kacha Birra	19	Zogoba and Hobicheka
Kedida Gamella	17	Sheshera and Holegeba
Omo Sheleko	20	Durgi and Mandoye
Total	104	11 (10.6 %)

5.3. Project Outcomes and Impacts

5.3.1. Preparation of a Handbook and Conducting In-service Training

As part of the baseline survey for the pilot project, the exiting textbooks for grades 5 – 8 were thoroughly examined by the project team to see the way in and extent to which issues related to natural resource management had been addressed. The analysis showed that most of the school subjects (particularly social studies) included topics related to environmental issues (SNNPR Education Bureau, 2001). The analysis of the textbooks also revealed a different fact that most of the topics related to natural resource management and environmental protection had been presented neither comprehensively nor systematically. Furthermore, environmental issues were not critically analysed or discussed. In other words, although the curricula had windows of opportunity to enable schools to address environmental issues posing challenges to life in the study areas and the nation at large, the presentation of the issues was neither comprehensive nor systematic. That was why a decision was made by the project office to produce supplementary teaching material (handbook) and train teachers to handle the issues effectively.

The handbook had five modules covering local, national and global environmental issues. The topics of the modules are indicated below:
- *Module One:* Energy and Environment
- *Module Two:* Population Growth vis-à-vis Use and Management of Natural Resources
- *Module Three:* Environmental Health and Personal Hygiene
- *Module Four:* Natural Resource Base of Ethiopia

– *Module Five:* Natural Resource Base of Kembata-Tembaro zone and Alaba Leyu *Wereda*

The people who wrote the modules were selected from a pool of highly competent professionals mainly from Addis Ababa University, Debub University, the Ethiopian Energy Authority and the Ethiopian Mapping Agency. In an effort to empower the teachers and school administrators of the selected schools, a five-day in-service training workshop was arranged for teachers, directors and supervisors, based on the handbook referred to above. A total of 115 educators (97 teachers, 11 directors and 7 school supervisors) attended the training. At the end of the workshop, participants prepared an action plan to mobilise the surrounding community to protect the environment, integrate environmental issues into daily lessons, start or strengthen environmental clubs and establish school nurseries.

Both teachers and directors of the respective schools substantiated that the handbook had been used with a high degree of enthusiasm. Teachers felt that it was comprehensive enough to accommodate the diverse interests of the different subject areas. The level of difficulty of the handbook was also considered appropriate and the language was rated as understandable (only one teacher in Mekalla School complained that some sections were difficult when dealt with individually). The pictures included in the handbook, which depicted diverse aspects of the region, were particularly appreciated by the users. The teachers said they felt more confident in teaching issues related to energy and environmental management, as the handbook provided them not only with detailed explanations, but also examples which were not available in the textbooks prepared by the Bureau of Education. The SIM Canada final evaluation report also considered the handbook "an excellent teacher's reference guide" (Paterson, 2008:29).

There were, however, some limitations observed in relation to the content and use of the handbook. The evaluation report referred to above indicates, for instance, that the handbook "provided relevant information at the local community level but little practical information as to what teachers and students could do to address these issues" (Paterson, 2008:29). The other challenge involved the high turnover of teachers who received the training (partly because of the unfavourable physical environment). In Sheshera School, for instance, only one out of the 11 trained teachers was still working in the school, as all the others had moved to different schools in the nearby *Weredas*. In Hobicheka as well, most of the teachers had transferred. The positive aspect of the challenge was that teachers who got transferred after the training continued using the information they gained, and some even established environmental protection clubs in the new schools, which were not part of the pilot project. Based on such experiences related to use of the handbook, the project schools strongly recommended that the handbook be disseminated to all the primary schools in the five *Weredas*. It was also suggested that key natural resources (e.g. Sinbita, Fulame, Shuppa and Lamo Forests), which were not covered in the handbook, be included during future revisions.

5.3.2. Establishment of Environmental Clubs

The Ethiopian Government's Plan for Accelerated and Sustained Development to End Poverty (PASDEP) duly recognises the role of environmental clubs in creating environmental awareness. The document emphasises that environmental awareness raising "will be directed at students through environmental clubs and at the public through the mass media and publication produced for the purpose" (MoFED, 2006:91). One of the major achievements of the pilot project is the establishment or strengthening of environmental clubs in all 11 schools. Table 5.2 indicates that eight out of the 11 clubs (72.7 per cent) were newly established following the in-service training programme.

In all the schools, club members included both teachers and students. They also contained both male and female students, although the proportion of the former was higher in all cases. There are indicators that the project managed to bring about positive changes in the attitude of club members. One such piece of evidence for this could be the practical action taken to rehabilitate degraded environments within and around school compounds. Following their establishment/revitalisation, most of the clubs prepared an annual plan of action and started operation accordingly.

The practical activities undertaken by the club members in their first year of operation included cleaning school compounds, preparing seed beds, demonstrating alternative energy technologies to the surrounding communities, preparing sites for tree planting and then planting trees in and outside school compound. The clubs mobilised their respective school communities and planted thousands of trees of different species within their school compound and/or degraded and abandoned farm and/or grazing lands around their schools.

For instance, the Kerekicho School established the largest school nursery with 20 nursery beds, where more than 63,000 seedlings of different species were prepared for planting. In the first year alone, the school managed to plant 13,929 seedlings in the school compound (Fig. 5.1) and on private holdings by distributing seedlings to some model farmers and all students of the school. Moreover, at the time of study, Kerekicho Primary School was found to be the cleanest school of all schools visited in the study. The school compound garden

Table 5.2: Environmental clubs' establishment and membership

Name of Kebele/village	No. of Teachers	Club Establishment and Membership						
		No. of students				Time of establishment		
		Male	Female	Total	% of female	Time		Remark
Durgie	All	24	16	40	40.0	End of 2006		Started after the training
Mandoye	3	45	10	55	18.2	End of 2006		Started after the training
Zogoba	3	24	11	35	31.4	End of 2006		Started after the training
Adancho	2	17	5	22	22.7	End of 2006		Started after the training
Hobicheka	15	55	45	100	45.0	End of 2006		Started after the training
Kerekicho	1	10	5	15	33.3	End of 2006		Started after the training
Sheshera	3	23	18	41	43.9	End of 2006		Started after the training
Holegeba Zato	7	67	33	100	33.0	Existed for years		Strengthened after the training
Gerema	13	80	33	123	26.8	Existed as agriculture club		Renamed and strengthened after the training
Mekalla	All	69	23	92	25.0	Existed as environmental development club		Renamed and strengthened after the training
Besheno	All	All	All			End of 2006		Started after the training

Project Outcomes and Impacts 89

Fig. 5.1: A teacher showing a tree planted in his school compound

was well managed. Indigenous tree species were planted in an orderly manner and kept well, flowers were grown around offices and staff rooms and they were regularly watered by students.

The pilot project invested a lot in the training of teachers, assuming that they in turn would train club members and others. No direct effort was, nevertheless, made by the project to build the capacity of school clubs and community leaders. With respect to sustaining the good work started by the environmental clubs, the participants of the focus group discussion underscored the need to train club members and to prepare and disseminate information brochures. It was also suggested that the inclusion of members of the training board and parent-teacher committees in the training programme would enhance replication of some of the 'good practices' in schools and communities currently not covered by the pilot project.

5.3.3. Development of School Nurseries

One of the aims of the pilot project was to encourage schools to start nurseries. As part of the package, some farm tools and seeds of different species were provided. At the time of the study, all 11 schools had started nurseries within their premises. A variety of seedlings were also sown, and most of the schools decided to plant indigenous tree species which were disappearing in the area. Multipurpose trees (that bear fruits and are used as shades) were selected by the environmental clubs. Additionally, the offices of agriculture in the nearby *Wereda* capitals agreed to supply seeds not easily accessible by schools. The project office, on its part, distributed 66 kg of seeds to the 11 schools.

It is worth noting here that the establishment/revitalisation of environmental clubs has been inextricably related to the development of school nurseries. Almost all the existing school nurseries were established following the in-service training of teachers.

5.3.4. Dissemination of Alternative Energy Technologies

Solar home systems (SHSs)
Many studies show that SHSs can, directly and indirectly, positively contribute to sustainable development (see Cabraal et al., 1996; Ahm, 1998; Kayastha, 2000). Direct monetary benefits of SHSs can include income generation[19] through the development of cottage industries, the creation of jobs, e.g. through installing and maintaining SHSs, and saving expenditure on batteries (dry cells), among others.

Studies also indicate a number of non-monetary benefits of SHSs (Cabraal et al., 1996; Ahm, 1998). Major examples include instantly available and higher-quality light, cleaner in-house air due to reduced or eliminated soot and fumes from open fire, kerosene and candles, improved safety due to reduced fire risks, or access to services such as TV, radio or fridges.

SHSs are particularly appropriate for community centers such as schools, churches and mosques in remote rural areas of less developed countries, such as Ethiopia, where there is no access to the energy grid. In consequence, all 11 schools of the pilot project were provided with a SHS (Fig.5.2). The SHS package included a panel, battery, charge regulator, inverter, lamp holder, energy saving bulbs, cables, transmission wire and switches. The initial purpose of providing

---

[19] A study carried out in Nepal has indicated that SHSs do not help to promote new kinds of income-generating activities. They do, however, assist already existing businesses to generate more income (Kayastha, 2000: 107).

Fig. 5.2: A solar home system being installed

the SHS was to supply schools with energy for lighting and operating radios/cassette players. The ultimate goal was, however, to create access to information about these alternative technologies, in communities where there is not such information, by using the schools as centres for information dissemination and skill development. All the schools were therefore highly encouraged to disseminate the information to their surrounding communities.

During the project, all schools except Durgie and Kerekicho, used the SHS for lighting and operating radios/cassette players. Access to electricity made it possible to use the evening hours more productively. Teachers used the light to prepare their lessons in the evenings. School board members used it to conduct board meetings in the evenings as board members were busy during daytime.

Before the installation of the solar panels, most of the schools had their school mini-media (mostly radios/cassette players) powered by dry cell batteries. The Besheno School, for instance, spent 21 Birr/week on average on the purchase of dry cell batteries alone. This accumulated to about 840 Birr in the academic year. One can thus easily calculate that the money spent for dry cell batteries alone over six-and-a-half years would be enough to buy a 40Wp SHS set. This gives evidence on the potential of SHS technology in Ethiopian schools once its initial costs are covered.

The attention of the community was attracted by the possibility to use the SHSs to charge mobile phones. Six of the 11 schools used the SHSs for this purpose. In Besheno, for example, the possession of mobile phones grew from less than 20 to more than 300 in a few months once the SHS was installed. There was network coverage in the region, but no access to energy before and people had to send their mobile phones to Kulito town – 38 km away – for recharging. The mobiles were used by farmers and traders to obtain market information and to contact lorry drivers when they wanted their products to be taken to the market. Almost all teachers and school committee members now own a mobile phone, which has, according to the school director, positively contributed to the effectiveness and efficiency of school administration and planning.

Immediately after the installation of the SHSs many communities asked to establish evening classes. Due to a scarcity of teachers, however, this could not be realised. In any case, the strong interest by the 'target group' can be seen as an indicator for the success of SHSs in rural schooling as part of a FBO-initiated development project.

Solar cookers and driers
Almost all people in the project areas use clay pots for cooking. In order to reduce their dependency on firewood and unhealthy smoke inhalation solar cookers and driers were supplied to all 11 project schools to test their appropriateness and potential for wider dissemination. Each of the solar cookers and driers cost 500 and 750 Birr respectively, excluding transportation. All schools made practical attempts to test the apparatus in their compounds and checked that the technologies were capable of providing the expected functions. Some of the schools also demonstrated the technology to the surrounding communities.

However, the participants of the focus group discussions appeared to be not very enthusiastic on the new solar cookers. Most of them felt that the solar cooker were unsuitable for their needs. Some complained about the relatively long time it took to cook the same amount of food. Other critics concerned the high dependence of this technology on solar radiation that did not allow to use it in the evenings and in the rainy season. Generally, in areas with no shortage of firewood, solar cookers seem to have only little potential for dissemination.

In general, people preferred the use of improved cooking stoves to solar cookers. The project tried to give comprehensive information about improved cooking stoves, known as Mirt in Amharic, promoted by the Ministry of Mines and Energy among others.

When properly used, the Mirt stove can reduce the firewood consumption by half. It is particularly constructed to prepare injera, the Ethiopian staple food. Focus group discussions showed that the interest to buy Mirt cookers is substantially high in the project schools

Project Outcomes and Impacts 93

Fig. 5.3: A solar cooker

Fig. 5.4: A solar drier

Fig. 5.5: An improved cooking stove called 'Mirt'

6. Case Study Two: Improving Access to Education

6.1. Background

> Of all the development activities that the church should participate in aggressively, education must stand out as a priority concern (Sirgiw Gelaw, 2007:33).

As discussed in Chapter Three, the policies of the Dergue regime adversely affected the scope of the EKHC to engage in educational activities for almost 20 years. Since 1991, there has been a significant move to revitalise EKHC's educational services. In March 2004, the participants of the EKHC General Assembly (the highest EKHC decision-making body) reviewed the then state of the Church's education ministry and passed a strong resolution calling for the revitalisation of the ministry.

A comprehensive survey was carried out in 2006 to determine the quantitative changes brought about by the EKHC Ministry and Leadership Transformation project (Horn and Tefera Talore, 2006). The finding of the survey indicates, amongst other things, that 27 per cent of the sample churches ran literacy programmes and only 3 per cent had first-cycle primary schools. In absolute figures, this would mean that there were 1,550 literacy programmes and 172 primary schools managed by the church throughout the country. The survey also revealed that 47 per cent of the schools were financed by local churches and 48 per cent supported by fees paid by parents.

On top of what has been done by congregations, the EKHC Head Office is running projects together with its European partners. These projects mainly aim at improving basic education in remote rural areas. One such project, which operates in four project sites in three Weredas in SNNPR and Oromiya region, was taken as a case example for this study. Accordingly 20 of the 73 education centres which were established by the project (five in each site) were visited during the field research phase of this study.

6.2. Key Objectives and Beneficiaries

The alternative basic education project was initiated and co-designed in 2004 by the EKHC and the Addis Ababa Office of the German Kindernothilfe (KNH). Implementation started in 2005 in three *Weredas* in SNNPR and Oromiya region. The management office is located in Alaba town. The project intentionally sought to operate in remote rural areas where road infrastructure and formal social and educational services were totally lacking. Most of the project areas were accessible only during dry season. The goal of the project was to increase sustainable community-based access to basic literacy programmes for children from poor and disadvantaged families, and to offer adult education services for remote areas. A pre-project survey showed that many parents in the project areas did not send their children to school. Instead children had to work in the household compounds or on farmlands and grazing fields.

The key activities of the project concerned the establishment of a targeted number of 80 education centres in the three *Weredas* by constructing low-cost rooms, the recruitment of 160 teachers (two per centre – one of the two was expected to be female), and the provision of educational materials for all centres at subsidised prices. The main beneficiaries of the project were children and functional literate adults who had no access to basic education. The targeted number of direct beneficiaries was 20,000, out of them 16,000 children and 4,000 adults. Beyond, the education committee at each school, *Wereda* education officers, and teachers were expected to benefit from the training and capacity-building activities of the project.

6.3. Major Achievements

The major precondition of the project was that the local communities had to do all the construction work of the education centres by themselves and to cover at least half of the construction costs on their own by supplying materials such as wood, stones and sand. Table 6.1 shows that this approach – which is not often exercised by FBOs and NGOs in Ethiopia – was quite successful with a total number of 73 education centres constructed in the four project sites.

Table 6.1: Education centre construction

Weredas	No. of functional centres	Structures and facilities				
		Concrete Foundation	Wall plastered	Doors and windows completed	Sufficient seats prepared	Toilets dug
Shashogo	20	3	17	13	10	5
Alaba/ Geremasite	19	4	18	15	13	1
Alaba	21	3	12	15	13	1
Siraro	13	2	6	7	7	0
Total	*73*	*12*	*51*	*50*	*43*	*7*
Percentage	91.3	16.4	69.9	68.5	58.9	9.6

Fig. 6.1: A typical basic education centre

What is not seen in Table 6.1 is the number of additional rooms constructed by the community. The project planned to establish education centres with two rooms but left it open for the communities to add more rooms if needed. This materialised in most cases. In Shashogo *Wereda*, for instance, almost all education centres now have more than two rooms – without additional financial or material support from the project. Education centres in Alaba *Wereda* and Gerema site are also following suit (see Fig. 6.2).

Fig. 6.2: Additional classrooms constructed fully by the community

By 2006/7, more than 12,000 children had been enrolled in the 73 centres (Table 6.2 and Figure 3), 45 per cent of whom were female. Shashgo *Wereda* had already exceeded its planned enrolment by 48 per cent, while Siraro was trailing far behind. Shashogo *Wereda* was able to accommodate enrolment beyond the planned number by constructing additional rooms. The *Wereda* assigned additional teachers in all the schools.

During the survey, a total of 154 teachers gave classes. All of them received in-service training at least once (Figure 6.4 and Table 6.3). All teachers, except one, hold the required qualifications for teaching. Table 6.3 shows that 54 per cent completed grade ten, 34 per cent grade twelve and 115 were TTI graduates. The community was supposed to cover part of the teachers' salaries. 60 per cent of the education centres paid their share continuously. Once again, Shasogo *Wereda* stands out with regard to pay teachers' salaries – with 95.0 per cent continuous payment. The worst performance was exhibited in Siraro Wereda with only 11.8 per cent. At this juncture, one must underline that demanding

Table 6.2: School Enrolment

Weredas	Level (2006/7 academic year)										% Female
	Level 1			Beginners			Total				
	M	F	T	M	F	T	M	F	T		Total
Shashogo	1,699	1,773	3,472	1,045	677	1,722	2,743	2,450	5,193		47.2
Alaba/Gerema	950	467	1,417	700	715	1,415	1,650	1,182	2,832		41.7
Alaba	460	433	893	1,258	1,070	2,328	1,818	1,460	3,278		44.5
Siraro	248	205	453	226	226	452	474	431	905		47.6
Total	3,357	2,878	6,235	3,229	2,688	5,917	6,685	5,523	12,208		45.2

payments for schooling services that are in Ethiopia otherwise often provided for free by FBOs and NGOs is a particular challenge.

Fig. 6.3: Children showed up even before the completion of the education centres

Table 6.3: Teachers' profile

Weredas	Level				Total	No. of centres which pay salary continuously	
	<Grade 10	Grade 10	Grade 12	TTI		No.	%
Shashogo	1	14	18	7	40	19	95.0
Gerema	-	28	10	-	38	10	52.6
Alaba	1	41	-	-	42	13	41.9
Siraro	-	-	24	10	34	2	11.8
Total	2	83	52	17	154	44	60.3
Percentage	1.3	53.9	33.8	11.0	100.0		

The project was implemented by the Evangelical FBO EKHC together with the German Kindernothilfe (KNH) in predominantly Muslim communities. One could therefore expect suspicion on the part of the beneficiaries. This was not the

Fig. 6.4: Teacher's training

case. The attempt by the project staff to strongly involve the community and local governmental officials appears to have achieved its intended goal; the degree of acceptance by the community and support given by the *Wereda* government officials attests to this fact. The collaboration between the Alaba *Wereda* government offices and the project management offices seemed exemplary.

Recently, the project has received extensive coverage through media (Ethiopian television). Mr Bedru, the Administrator of Alaba *Wereda*, publicly acknowledged that the project brought significant positive change in the educational landscape of his *Wereda*.

Most of the education centres were established on farmland or grazing fields adjacent to villages with no other formal social or educational services. Since their establishment, some of the educational centres attracted the establishment of other services like farmers' training centres or small medical centers. These considerable 'side benefits' contributed to the gradual establishment of 'comprehensive development hubs' in the project sites.

The Keranso centre is one of the 19 education centres in the Gerema site, located in Badewacho *Wereda*. It has two features which attract the attention of any casual visitor. The first is the high number of older children – aged ten and above – coming to the centre and reciting the alphabet and numbers for the first time in their life. The second feature is the extraordinary positive performance of girls. Girls took every first, second and third position in performance rankings in the 2006/7 academic year. This could by used as a showcase for families with negative attitudes towards girls' education.

The concept of the project was based on the mobilisation of local beneficiaries and their own material and financial resources and on mostly self-made low-cost construction of education centres. This was, obviously, not what some of the targeted communities expect FBOs and NGOs to do. According to the EKHC project officers, the most difficult challenge faced by the project was, therefore, related to the mobilisation of own resources of the beneficiaries. Some government officials openly accused the project to 'overburden the community' by demanding construction materials for the schools and salary for the teachers. Particularly in the beginning of the project the project officers were frequently asked either to cover all expenses or to terminate the whole project.

Fully externally funded development projects are likely to inject a sense of dependency to local beneficiaries, especially in remote rural areas. . According to the EKHC project officers, a 'dependency syndrome' is found in rural Ethiopia particularly in areas where the government's Productive Safety Net Programmes are implemented since 2005. The EKHC-KNH alternative basic education project faced difficulties to find people working for free to build the educational centres especially in areas in which other development projects/programmes provided income generation opportunities through cash-for-work approaches.

7. Case Study Three: Help a Child

7.1. Background

The plight of Ethiopian children
In Ethiopia, about 47 per cent of children less than five years are chronically malnourished, and 24 per cent are severely stunted (EEA, 2008). The same source indicates that 38 per cent are underweight (low weight for their age) and 11 per cent wasted (low weight for their height). The levels of stunting, low weight and wasting are higher among rural children than for their urban counterparts. A comparison of data from 2005 with that of 2000 shows some improvement in the nutritional status of children during this time period (EEA, 2008). The percentage of stunted children fell by 5 per cent from 52 per cent in 2000 to 47 per cent in 2005. Similarly, the percentage of underweight children declined by 19 per cent from 47 per cent in 2000 to 38 per cent in 2005. There was, however, no change over the five-year period in the percentage of wasted children. In the country as a whole, the mortality rate of children under 5 years is 172 per 1,000 live births, and a large number of infants have low birth weight. About one million children in Ethiopia have lost their mothers or both parents due to HIV/AIDS.

Most children in the country are engaged in productive and household activities such as herding cattle, weeding, harvesting, taking care of siblings and various household chores. Working children mostly lack access to occupational safety, work for long hours, get very low wages and work in an often dangerous environment. The situation of children in especially difficult circumstances, such as street children, HIV/AIDS orphans, children with disabilities and juvenile delinquents, seems even worse. It is known that there are over 100,000 children working and living on the streets in Ethiopia (MOLSA, 2002). Family poverty, family disintegration, abuse and neglect by parents, lack of educational opportunity and the social environment (peer influence) are the major causes of what is known as 'streetism.' Some of these street children are involved in theft, drug addiction, violent acts and various offensive behaviours. The relationship with the dwellers of the urban centres is therefore not that positive in most urban centres. A study conducted in 2002 by the Ministry of Labour and Social Affairs (MOLSA) indicated that there were about one million HIV/AIDS orphans in Ethiopia, and the number was estimated to rise to 1.8 million by 2010.

International and national efforts
The 27th General Assembly of the UN held a special session (in May 2002) and adopted a resolution regarding "a world fit for children" (Unicef, 2005). The priority areas identified were: promoting healthy lives, providing quality education, protecting against abuse, exploitation and violence; and combating HIV/AIDS. A joint report by UNAIDS, Unicef and USAID, referred to as 'Children on the Brink 2002,' outlined key strategies for supporting orphans and other children affected by HIV/AIDS. The strategies, which would focus on helping families and communities cope with the crisis, were adopted by local, national and international groups (UNAIDS, Unicef and USAID, 2003). The strategies form a framework for interventions to match the massive scale and lengthy duration of this unprecedented crisis. The five strategies are as follows:
– to strengthen and support the capacity of families to protect and care for their children,
– to mobilise and strengthen community-based responses,
– to strengthen the capacity of children and young people to meet their own needs,
– to ensure that governments develop appropriate policies, including legal and programmatic frameworks, as well as essential services for the most vulnerable children, and
– to raise awareness within societies to create an environment that enables support for children affected by HIV/AIDS.

What is more, key principles have been jointly developed by the aforementioned organisations to guide programmes aimed at mitigating the effects of HIV/AIDS on children and their families. The principles provide practical guidance for implementing the five strategies presented above, represent a common point of reference for the various groups working to help children, families and communities and can help guide collaborative action at all levels – local, national, regional and global. The 12 principles developed by UNAIDS, Unicef and USAID namely:
– to strengthen the protection and care of orphans and other vulnerable children within their extended families and communities,
– to strengthen the economic coping capacities of families and communities,
– to enhance the capacity of families and communities to respond to the psychosocial needs of orphans, vulnerable children and their caregivers,

- to link HIV / AIDS prevention activities, care and support, for people living with HIV / AIDS, and efforts to support orphans and other vulnerable children,
- to focus on the most vulnerable children and communities, not only those orphaned by HIV / AIDS,
- to give particular attention to the roles of boys and girls, and men and women, and address gender discrimination,
- to ensure the full involvement of young people as part of the solution,
- to strengthen schools and ensure access to education,
- to reduce stigma and discrimination,
- to accelerate learning and information exchange,
- to strengthen partners and partnerships at all levels and coalition among key stakeholders, and
- to ensure that external support strengthens and does not undermine community initiatives and motivation.

The Ethiopian Ministry of Labour and Social Affairs has prepared a National Plan of Action for Children (2003 – 2010). As a background to the action plan, the current plight of the country's children has been well studied and documented. The document indicates that about 52 per cent of Ethiopia's population is below 18 years, while those below 15 years amount to 44 per cent. The projected number of primary school-age children (ages 7 – 14) was 17.71 million and the number of secondary school-age children (age 15 – 18) about 6.8 million in 2010. According to official sources, net primary school enrolment stood at only 30 per cent in 2002.

The Ethiopian national plan of action focuses on the themes on which the UN's special session on children agreed. The following are the goals set for the period from 2003 to 2010.

- *Promoting healthy lives:* This involves increasing health care coverage to 62 per cent, reducing maternal and child mortality by one-third, improving nutrition, sanitation and water facilities and controlling the major killer diseases such as HIV / AIDS, malaria and TB.
- *Providing quality education:* This involves the expansion of early childhood education, providing quality primary education to 90 per cent of the Ethiopian children, improvement in the quality of teachers, allocation of larger amounts of resource (budget) and working towards narrowing the gaps between regions and the sexes. Special assistance is planned to be provided to children in pastoralist areas and those with disabilities.
- *Protecting children from abuse, exploitation and violence:* This involves the registration of children at birth, a revision of laws, raising awareness about harmful traditional practices and improving the juvenile justice system.

Fig. 7.1: Children in a village basic education centre, southern Ethiopia

7.2. EKHC Children's Services

Following the consolidation of development services and introduction of externally supported development projects centrally coordinated by its head office, the EKHC started activities aimed at addressing the diverse needs of children in general, and orphans and vulnerable children in particular. The EKHC started these services through its childcare projects mainly funded by World Vision Ethiopia and Red een Kind from The Netherlands. Other donors of these projects included the Christian Children's Fund of Canada (CCFC) and Hope International. Services provided for children and their families included education and health services, the construction of roads and bridges as well as the provision of oxen, farm equipment, seeds, fertilizers, clothes and educational materials.

7.3. The Kuriftu Children's Home

Red een Kind (ReK) is one of the EKHC's main European partners, with a history of more than two decades of strong partnership. The vision of ReK is to "help children who have little or no hope of a decent life" (www.redeenkind.nl). ReK focuses on children in less developed countries who are subject to poverty, discrimination, diseases or natural disasters. ReK supports children irrespective of their religion, caste, colour, race, creed or gender.

The partnership between the EKHC and ReK was initiated by Dr. Mulatu Bafa, the then General Secretary of the EKHC and Mrs. Anky Rookmaaker, the then Secretary of ReK, in 1987. The first joint project was the establishment of a children's home associated with the famine that shocked Ethiopia and the world in the mid-1980 s. This is shown in a letter sent by Mrs. Anky Rookmaaker to Dr Mulatu Bafa on December 28, 1987.

> With all the publicity about the famine in Ethiopia there is much interest to give for children in your country. This means that we probably can now get the necessary funds for such a children's home, if we get the necessary information from you. Therefore let us have now as soon as possible the necessary budget and other information for starting this home.

In 1988 the EKHC acquired a vast amount of land (roughly estimated to be 50 ha) for the establishment of a children's home in Kuriftu, Oromiya region, not in Addis Ababa as initially planned. The Kuriftu site was "chosen by the help and personal presence of the Children Commission Commissioner, Comrade Teserawork" (letter written by Dr Mulatu Bafa to Mrs. Anky Rookmaaker on 21 July, 1988). However, it took three years to bring the first children to their new home. This apparently created a great deal of unease on the part of ReK, as can be seen from the following statement (letter sent by Mrs. Anky Rookmaaker to Dr Mulatu Bafa on May 21, 1991):

> It is now May 1991 and I have not heard about the children. Every evening on TV we see the undernourished and dying children of Ethiopia, and it is hard to explain to our people that no children can be helped yet. They gave very generously for the buildings and do not understand the delay. Please explain.

After years of ups and downs, the children's home in Kuriftu was finally opened at the end of 1991. Its aim was to address the educational, economic and psychological needs of children who had lost their parents during the 1984/85 famine (Daniel W/Giorgis, 2003). Initially the centre gave home to 89 orphans (49 boys and 40 girls) from different regions of the country: 20 from Oromiya, 46 from Addis Ababa, 5 from Amhara, 5 from Tigray, 8 from SNNPR and 5 from Harari. Five of them were later adopted and went overseas, one was expelled from the children's home for disciplinary reasons.

Table 7.1: Educational level of the children at Kuriftu Home (as of 2003)

Grade	Age			Male	Female	Total
	10–14	15–19	>20			
1–4	1	-	-	1	-	1
5–6	1	2	-	2	1	3
7–8	10	10	-	10	10	20
9–10	1	26	-	8	19	27
Preparatory	-	3		2	1	3
Vocational			9	9	-	9
Collage		2		2	-	2
Degree			18	9	9	18
Total	**13**	**43**	**27**	**43**	**40**	**83**

Source: Daniel W / Giorgis (2003)

More recent information from June 2009 shows that 50 children from the Kuriftu children's home have graduated with certificates, diplomas or degrees (Table 7.2; Figure 7.2), one-third of the graduates specialised in auto and general mechanics.

Table 7.2: Children graduated from colleges as of 2009

S/n	Area of specialisation	Level completed	Male	Female	Total
1	Catering	Certificate	1	10	11
2	Hair dressing	Certificate	2	6	8
3	Auto & general mechanics	Diploma	18	0	18
4	Accounting	Diploma	2	1	3
5	Secretarial science & office management	Diploma	0	5	5
6	Marketing & business management	BA	1	1	2
7	Nursing	BSC	1	0	1
8	Hydraulic engineering	BSC	1	0	1
9	Law	Diploma	1	0	1
Total			**27**	**23**	**50**

However, in 2009 only 10 per cent of the children have finally obtained college degrees. Only one of the female children managed to get a BA / BSC degree. Given the relatively good care these children received over years in the orphans' home these figures are not satisfying at all. An in-depth analysis of its underlying reasons is hence recommended.

Fig. 7.2: Some of the students at Kuriftu centre celebrating the graduations of their long-time friends

7.4. Community-Based Child Sponsorship

Community-Based Child Sponsorship (CBCS) Programmes allow private sponsors to pay for individual children. Sponsors have options to choose from three different programmes. First, they can support a child whose parents or guardians are very poor and live at home or with relatives. The child in this case is provided with good meals on a daily basis, as well as readily available medical care. He/she also goes to school and participates in a programme of educational support. Second, sponsors support a child that already attends school, but belongs to a deprived group and is at great risk of leaving school in order to supplement the family income. The child in this case is sponsored for one meal per day and educational support. Third, sponsors give an entire (foster) family a new start. Members of the family participate in programmes aimed at strengthening self-sufficiency while the children pursue education. Extra food aid, clothing and medical care are provided temporarily.

The EKHC has decades' experience with CBCS programmes throughout Ethiopia with partners mainly from Europe and North America. As of 2003, in Oromiya region a total of 6,293 children were supported by ReK-sponsored

CBCS programmes (Daniel W/Giorgis, 2003). A recent report issued by the project management office at Kuriftu indicates that most of the beneficiaries of CBCS programmes were children in grades two to five, with the boy-girl ratio almost equal (Table 7.3).

Table 7.3: Number and grade level of children supported through ReK in Babogaya

Grade	Male	Female	Total
KG	12	18	30
1	45	30	75
2	85	93	178
3	118	92	210
4	151	165	316
5	132	149	281
6	35	32	67
7	17	14	31
8	6	4	10
9	3	2	5
Total	604	599	1,203

7.5. A Turning Point in the EKHC-ReK Partnership

The EKHC considers ReK one of its strategic partners in Europe, not only because the partnership has withstood 20 years of ups and downs, but also because of the keen interest of ReK to scale up the partnership in terms of both programme diversity and amount of resources allocated to each of these programmes. At present, new projects have been started in the areas of basic education, vocational training and the prevention of HIV/AIDS. Only one of these newly initiated projects, namely basic education project, is described in the next subsection, just to show how the partnership grew from a single children's home meant to serve less than 100 children into a plan to establish 35 fully-fledged first-cycle primary schools with a capacity to enrol 14,000 children from communities with no access to any kind of education.

The project initiation
The basic education project was initiated and designed by the EKHC in 2006 and supported by ReK. The goal of this project was to increase sustainable community-based access to basic education for children from poor and disadvantaged families, and offer adult education services in those areas relatively least privileged. The project had three overarching aims, namely: direct poverty reduction, strengthening civil society and lobby and advocacy.

The following are the activities planned to archive these aims.

Activities related to 'direct poverty reduction'
– Establishing 35 primary schools.
– Assigning four teachers to each school.
– Designing/adapting co-curricular activities.
– Initiating and running adult functional literacy in 10 schools.
– Preparing post-literacy reading materials for adults on issues such as environmental protection, HIV/AIDS, food security, hygiene and sanitation, eradication of harmful traditional practices, family planning and life skills.
– Providing basic and supplementary educational materials for all the schools.
– Undertaking experience sharing visits to areas/farms to facilitate learning from the success of others. Such a tour is also planned for primary school teachers across the project *Weredas* so that one community benefits from the 'good practices' of other communities.

Activities related to 'strengthening civil society'
– Organising and conducting a series of training workshops on issues related to basic education for civil society organisations in the project area.
– Initiating the idea to establish a network among government, civil society, and

non-governmental organisations actively engaged in the education of children in disadvantaged areas.
- Encouraging religious leaders to use the materials prepared for adult education as part of their religious education (tools for wholistic ministry).

Activities related to 'lobby and advocacy'
- Offering training that focuses on the value of education and mobilisation of the community to participate in school construction.
- Using public places like schools, churches and mosques to disseminate information by using religious leaders as promoters of the idea of 'education for all.'
- Preparing awards to encourage girls who demonstrate outstanding performance, as well as their parents.

Preliminary achievements
The basic education project was in its third year when this study was undertaken. According to an official report of the project office, the super structure work, dry latrine excavation and masonry work for 18 of the 35 basic education centres (51.4 per cent) was completed. 99 per cent of the construction materials used were locally supplied. What seems more remarkable in this instance is that thousands of children have already started classes in makeshift classrooms prepared mostly by local communities.

At the beginning of the first semester of the 2008 / 2009 academic year, a total of 5,769 children (2,962 males and 2,807 females) were registered in the centres found in the SNNP and Oromiya regions, and 5,367 had successfully completed their first semester's education. Of the registered children, 452 (230 males and 222 females), or 7.8 per cent of the total, dropped out of school for different reasons. The reports from respective district project offices indicate that early marriage, family mobility and sickness were among the main causes for dropout.

A mid-term evaluation of the project conducted by an independent team of consultants also confirmed that it was on the right track: "[W]ith about one-and-a-half years to go before the project ends it is the opinion of the evaluators that the main objectives regarding quality access to education can be achieved. Access to education has even exceeded the targets (numbers) as formulated by the partner organisation" (Poyck, 2009:5). The evaluation team also observed problems in the quality of the schools: "[I]n none of the schools visited are there proper sanitation facilities, and there is no water. Potable water is unavailable in schools. Sometimes there is no electricity, there is no fence around the schoolyards, there are no sports facilities and there are no posters or banners or otherwise child-friendly and colourful teaching aids".

Table 7.4: Enrolment by region

S/N	Region	Grades	Registered			Dropouts			1st Sem. Complete		
			M	F	T	M	F	T	M	F	T
1	SNNPR	1	1,226	1,129	2,355	124	97	221	1,128	1,062	2,190
		2	898	867	1,765	48	64	112	850	797	1,647
2	OROMIYA	1	324	287	611	28	28	56	296	259	555
		2	514	524	1038	30	33	63	484	491	975
Total			2,962	2,807	5,769	230	222	452	2,758	2,609	5,367

Fig. 7.3: Girls enrolled in one of the basic education centres in Alaba *Wereda*, SNNPR

7.6. Concluding Remarks

The two-decade-long cooperation between the EKHC and ReK was characterised by many ups and downs. This makes it a good example for development-concerned south-north partnerships of FBOs and other NGOs. Let us mention three of the landmarks in the relationship between these two organisations as a way of concluding the chapter. First, the high initial level of enthusiasm of ReK to support children in Ethiopia. Second, flexibility was demonstrated in their relationships. The children's home was originally planned to be established in Addis Ababa, but this idea was changed and the centre eventually established at Kuriftu. Third, the two organisations agreed to scale up both the type and magnitude of their interventions, thereby creating opportunities for tens of thousands of children and adults in Ethiopia to get themselves out of material, social and psychological scourges of poverty.

8. Case Study Four: Transforming Theological Schooling

8.1. Background

There is a growing interest among Ethiopian FBOs to focus (or refocus) their services on the use of locally available human and natural resources, and building the capacity of people to render not only a spiritual service, but also physical and social services. Theological schools have a pivotal position in the overall management of church-based services, and hence play a major role in the efforts to reorient the existing, largely one-sided, programmes to more wholistic services. A study conducted in Pakistan revealed that a leadership crisis emerged in the Church of Pakistan because of "poor training programmes to produce leaders" (Sultan, 2001:333). The same study further noted that "many development activities initiated by the bishops were based on predefined theological convictions which overlooked the socio-political realities faced by people at a grassroots level".

The EKHC runs one degree-granting college, five diploma-granting colleges and more than 200 Bible schools throughout the country. The analysis of their curricula reveal that the socio-economic development components of the EKHC's mission are not sufficiently addressed as the curricula appear to have been designed to equip the graduates mainly with spiritual and mental knowledge and skills.

A continuous effort had thus been made over the past six years by the EKHC central office to integrate development education and questions of environmental sustainability into the curricula of the EKHC theological institutions. This chapter shows these efforts and their resulting outcomes.

8.2. An Untapped Resource

Hundreds of theological institutions are owned and run by the Ethiopian Evangelical churches. It has long been substantiated that theological institutions in general, and rural Bible schools in particular, can be centres of transformational development that addresses both socio-economic and spiritual needs of community members (Batchelor, 1993). To this end, theological institutions can pursue two approaches. The first involves offering short-term courses and organising seminars aimed at creating awareness and developing skills on issues that enhance community development. Additionally, they can design a fully-fledged course on development education and make it part of the Bible school's training programme.

The second approach, integrating a course or courses on community development into the Bible schools' curricula, would give prospective church leaders and other staff a chance to carry out a more rigorous analysis and interpretation of concerned issues. Such a course is expected to balance academic knowledge with theological discourse and theory with practice. The success of the course, or courses, depends to a great extent on the ability to develop practical skills that improve the lives of people in a given community. This, in turn, calls for designing the course / courses based on a systematic analysis of specific situations and needs. It is also important to note here that the second approach succeeds only if pre-service or in-service training of Bible school teachers is made part of the programme. The major aim of the EKHC's efforts to integrate development education into the curricula of theological institutions was thus to build the capacity of theologians to plan and implement community development activities in their local churches, and carry out development education in Bible schools and higher level theological institutions throughout the country.

8.3. Situation before the Integration Efforts

As indicated in Chapter Three, most of the FBOs in Ethiopia seem to have a high degree of commitment to render what they call a 'wholistic service,' i.e. serving the physical, social and spiritual needs of an individual. The EKHC's defines its own mission as the proclamation of *"the good news of Jesus Christ and His kingdom to the peoples of Ethiopia and beyond so that they receive eternal life, become Christ's disciples and are fulfilled spiritually, socially, mentally and physically to the glory of God"*. In order to fulfill this mission EKHC uses a number of strategies. Formal theological education is one of them.

The theological colleges and Bible schools are among some of the instruments used by the EKHC to meet, as mentioned above, the spiritual, social, mental and

physical needs of people in Ethiopia. The more specific objectives of the EKHC Bible schools are namely:
- to produce Evangelists, missionaries, pastors, teachers, leaders and Bible interpreters who have deep knowledge of the word of God,
- to develop the spiritual, social and mental skills of ministers, and
- to strengthen the Great Commission through enhancement of gospel outreaches, church planting and promoting discipleship.

It is thus clear from the above paragraphs that the EKHC follows a wholistic mission and the theological institutions in general, and Bible schools in particular, are expected to help the EKHC to fulfill this mission. In 2002, an EKHC-internal study was conducted to find out how far the Bible schools' curricula were designed in such a way that they enabled the EKHC to '*develop the spiritual, social and mental skills* of the trainees,' as stated in the objectives. The assessment of the curricula revealed that the courses being offered in the higher Bible schools were mostly theological. In fact, there was only one two credit hour course on 'English language' and another two credit hour course on 'family management.' All the other courses dealt with one or the other components of theological education.

A closer analysis of the titles of the courses revealed that courses related to community development had been given only a marginal position. Furthermore, the basic preconditions to teach such courses in a qualified way were not given: teachers at all levels were not properly trained to address issues related to community development, and materials necessary for development education were not available. Based on the findings, it was thus recommended that a course be developed on "Development Education" and fully integrated into the curriculum. The major aim of such a course, it was proposed, would be to enable the graduates of the Bible schools to understand not only the development issues of the contemporary world, but also make a significant contribution to the fight against poverty and food insecurity in their respective localities.

8.4. Integration of Development Education

The response of EKHC management to the recommendation of the aforementioned study was quick and positive. The idea was accepted and the church's Capacity Building and Community Empowerment Programme (CaBCEP) and Bible Schools Coordinating Office took the responsibility of developing the course syllabus, and offered the course first to directors and then to other teachers from higher level Amharic Bible schools. The following were the specific course objectives:

- to acquaint trainees with both secular and biblical perspectives on development,
- to create awareness on the general state of development in Ethiopia as compared with other counties in Africa and in the world,
- to create awareness of the relationship between environmental care and overall development,
- to establish the biblical basis for the church's involvement in environmental management and protection,
- to identify diverse views regarding the relationship between gender, population and development,
- to distinguish different perspectives on the causes of, and solutions to, the problems of poverty and food insecurity, and
- to enable students to prepare practical projects aimed at bringing about positive changes in their own communities.

The actual training of higher level Bible school directors and teachers was conducted three times, and a total of 146 participants attended. A number of educational materials including supplementary books and modules were prepared and distributed to all the Bible schools (Table 8.1). The training sessions and teaching material preparation were financed by the Evangelischer Entwicklungsdienst (EED), Germany, one of the main and longtime partners of the EKHC. On top of the training manuals, a sourcebook entitled "Perspective on Poverty and Development" was prepared and published in 2006 with financial support from Samaritan Purse, Canada. The sourcebook is now being used as one of the essential readers for Bible schools and theological colleges throughout the country (mainly by the Community Development Ministry Major, Evangelical Theological College and the Leadership and Management Programme, Mekaneyesus Seminary).

Table 8.1: Number of directors and teachers trained and teaching materials prepared

S/n	Activities	No.
1	Directors and teachers trained in development education	103
2	Directors and teachers trained in natural resource conservation and management	43
3	Teaching material on development education distributed	650
4	Teaching material on natural resource conservation and management distributed	100

What appears to be more encouraging is that almost all higher level Bible schools have now fully integrated the two courses related to development education into their curricula. The interim evaluation also indicated that the teachers who took part in the training were handling the courses very well and using the practical skills to change their own lives and those of the surrounding community. The higher level Amharic Bible school directors who took part in the Durame awareness creation and curriculum development workshop, as well as Langano, Yirga Chefe and Awassa training on development education, were found to have been trying hard to integrate development education into their respective schools. Furthermore, most of the trained directors and teachers were actively engaged in practical activities at household and community levels. Five cases out of many are presented here to demonstrate that it is possible to "not only preach but also practice community development". The following is based on information gathered during a focus group discussion with Bible school directors and an interim evaluation done by EKHC CaBCEP staff.

A. Shamena

- Trained 17 students on development education and natural resource management
- Started compost making for organic grain production
- Started keeping poultry to generate income
- Started horticulture for own consumption and income generation
- Trained Evangelists on sustainable development (which shows a significant change in attitude – previously "Evangelists had to do only Evangelistic work")
- Started rainwater harvesting from roofs (income in one year: 5,500 Birr from the EKHC District Office and 6,000 Birr from six congregations)

B. Boditi

- Hired a full-time development worker
- Trained 10 students on development education (extension classes attended mainly by church '*ministers*' and government workers)
- Established a development committee at Bodditi local church and gave training on community development for 10 more local churches in the surrounding areas. As a result, the Boditi local church started turning 'green' with vegetables and fruits and was used as a demonstration plot for other churches
- Started horticulture – ten households followed suit after getting advice and encouragement
- Completed preparations for an apiculture demonstration site

C.   Chencha

- Trained 37 students on development education
- Church staff received considerable incomes by selling apple, plum and medicinal plants (one reported earnings of 11,000 Birr in one year)
- Compost making and using became a day to day routine
- Two local churches received training on development and established a development committee. The local churches earned 20,000 Birr per year from selling fruits and vegetables. They were planning to increase this income to 30,000 Birr in future years.

D.   Amburse

- Trained 37 students on development education. Training was cascaded to six junior Bible schools and 22 local churches
- Started an initiative to use produce compost
- Started a literacy programme with 150 children from particularly poor families. A new teacher was hired for this purpose.
- Established a horticulture demonstration centre in the school compound which also acted as a source of income for the school
- Encouraged female students to start income-generating activities like embroidery, poultry, sheep raising, or apiculture

E.   Wolaitta Sodo

- More than 1000 students from junior and higher level Bible schools took part in development education training
- 21 hand-dug water wells and 33 pit latrines constructed as a result of mobilisation by the students who participated in development education trainings
- 80 traditional and modern beehives produced
- Students started several small-scale businesses like retail shops, grain mills, or cattle fattening
- Two kindergartens started

8.5. Community Development at ETC

The Evangelical Theological College (ETC) is owned by the EKHC. It runs its services in tandem with the International Evangelical Church (IEC), Serving in Mission (SIM) and the Evangelical Churches Fellowship of Ethiopia (ECFE). The main aim of the ETC is to educate highly qualified church ministers by equipping 'their head' (providing instruction in biblical, theological and human development disciplines), 'their heart' (providing nurture for Christian character) and 'their hands' (providing skills training for more effective services).

The start of the Community Development Programme at ETC was interesting. In the middle of the effort to integrate development education into the curricula for higher level Bible schools, the EKHC organised an informal consultation meeting. The participants of the meeting were the EKHC Deputy General Secretary for Development Programmes, the ETC Principal and Dean, the Tear Fund UK Regional Advisor, the Head of the EKHC Medical Ministries Department and the Head of the CaBCEP. They together proposed to integrate issues related to community development and HIV/AIDS into the ETC curricula. The CaBCEP Head had copies of the syllabus prepared for the Bible schools and distributed them to the participants of the meeting. The ETC Principal and Dean expressed their willingness to offer a course on Development Education at ETC.

A Development Education course was then given to graduating class students taking the degree programme as an optional course. Participants were church workers representing most of the denominations under Evangelical churches of Ethiopia. At the end of the first semester, the course participants invited the ETC Principal to their classroom and expressed that "the course is very helpful for our services and should therefore be made a core course, not optional" and "effort must be made to start community development as a major programme". Mr Simeon Mulatu, the then ETC Principal, agreed to work in this direction.

Taking the requests of the students into consideration, Development Education was soon offered as a core course at ETC. In the following the ETC management and EKHC CaBCEP continued working on a curriculum package to include community development as a major programme at ETC. The package, having six core courses and a number of electives, was developed and a Community Development Major Programme officially launched as of January 2005. This marked a turning point in the history of theological education in Ethiopia, because for the first time a programme which focused on building the capacity required for transformational development had been officially launched at 'major programme' level. The Community Development Major Programme offers the following core courses and electives:

Core courses	Electives (4 courses to be chosen)
– Perspectives on poverty and development – Population dynamics and development – Natural resources conservation and management – Special topics in community development – Appropriate technology and poverty alleviation – Planning, implementing and managing community development projects	– Food security strategies – Education for sustainable development – Community mobilisation methods and tools – Participatory action research – Community development practicum – Church and society – Urban ministry – The response of the church to HIV / AIDS in Ethiopia

The new major programme enjoyed a high level of acceptance by students from the very beginning. More than 40 students (about 40 per cent of the total enrolment for one year) showed interest in the first registration. These students came from all major denominations of the Ethiopian Evangelical churches – people educated to serve in leading positions in the churches, to become lecturers in theological institutions and to manage social and development services in their respective denominations. The Community Development Major Programme has to date graduated about 60 students in three batches.

8.6. Concluding Remarks

The case study presented in this chapter shows the progress already been made with respect to the integration of sustainable development related issues in educational programmes of the EKHC. In general, the efforts made and achievements witnessed in EKHC Bible schools and colleges over the last six years show that theological institutions can be catalysts for knowledge transfer and awareness raising on sustainable development in Ethiopia. How far these schools and colleges can substantially act as centres of positive socio-ecological transformation, however, highly depends on their vocational and entrepreneurial skills. The Community Development Major Programme is already playing a key role in Ethiopia in producing highly qualified graduates with the skills and knowledge to better integrate sustainable development issues into the curricula of theological institutions.

Concluding Remarks

Fig. 8.1: Community Development Major Programme students attend a class on improved stoves

Fig. 8.2: Community development students demonstrating the use of an improved cooking stove

Fig. 8.3: Community development students demonstrating the use of a bio-sand water filter

9. Concluding Summary and Way Forward

9.1. Concluding Summary

Similar to other African countries, the engagement of faith-based organizations (FBOs) in the development sector in Ethiopia is tremendous and manifold. Most Ethiopians are Sunnite Muslims, Ethiopian Orthodox, Protestant, or Catholic Christians. Many are strongly religious. This historically gave FBOs in Ethiopia solid grassroot foundations and strong participatory connation. A wide diversity of different Christian and Muslim FBOs with diverse backgrounds and interests work throughout the country – from Addis Ababa downtown to the remotest villages in the Southern Region. Although a fraction of civil society and NGO movements, FBOs are, however, still insufficiently integrated in the development discourse and often operate under the radar of international and multilateral donors. This is also the case in Ethiopia. It can be to a large extent reasoned by the fact that very little empirical research has hitherto been undertaken to assess the scope and impact of FBO activities in Ethiopia. At the same time, some scholars express scepticism on the underlying motives of FBOs to engage in development activities at all and raise questions on how far they are or should be integral parts of the FBOs mission.

Upon this background, this study aims to contribute to a debate among practitioners and researchers on the actual and potential contribution of FBOs to development in Ethiopia. This is exemplified by the work of the Ethiopian Kale Heywet Church (EKHC) in rural areas of southern Ethiopia in thematic fields of schooling, ecological balance and food security. Cross-sectional and case study designs form the principal approaches used in the study, and both quantitative and qualitative methods were used.

FBOs established the first 'modern' civil society organizational structures in Ethiopia as early as in the 1930s. The focus was put on grassroot, rather isolated and small scale actions. During the Dergue regime, most FBOs were forcibly expropriated and their work massively restricted. After the downfall of the Dergue in 1991, there has been a significant move to revitalise FBOs. Since the mid-1990s,

FBOs development related engagement underwent a renaissance in which they increasingly shifted their attention away from isolated small-scale projects and relief and rehabilitation programmes towards larger scale, more professionalised, often donor-co-funded and more integrated development programmes.

With regard to the EKHC, this research report illustrates the exceptionally strong human, technical and institutional capacities and social mobilization networks that larger FBOs have established locally in rural Ethiopia. The EKHC has a total of 5.7 million members in Ethiopia (as of 2006), organized in around 6000 local churches. Between 1991 and 2006, the total number of EKHC members increased by 25.5 per cent. Similar to this trend, the development programmes of the EKHC has grown in terms of both programme diversity and coverage. This particularly applies for Southern Ethiopia, namely SNNPR and Oromiya regions. In 2006, 27.2 per cent of the local EKHC churches are engaged in literacy programmes; 42.5 per cent in some form of development programmes, mostly focusing on pro-poor income generation. The list of education and development related activities of the 2007–2012 strategic plan reads impressive: it foresees activities to strengthen literacy programmes in 3,000 local churches; to establish 95 new primary schools, 83 adult education training centres and one teacher training institute; to create job opportunities for 300,000 women and youths by engaging them in income-generating schemes; among many others.

The findings of this study demonstrate that education continues to be at the core focus of EKHC's activities in Ethiopia. The contribution of the EKHC to rural schooling is relatively strongest in far remote rural areas of southern Ethiopia. In these areas the EKHC systematically targets population groups that were previously not or inadequately served with educational services for one reason or another. However, the whole theological educational system of the EKHC is currently undergoing a transformation process towards more integrated schooling. For example, the Evangelical Theological College (ETC) of the EKHC launched a Community Development Major Programme in 2005. This marked a turning point in the history of Evangelic theological education in Ethiopia, because for the first time a 'major' level programme focused on integrated capacity building with the aim to contribute towards community-based sustainable development. Graduates of the Community Development Major Programme were found to play important roles to integrate development and environmental issues into the curricula of rural schools across the country.

The survey conducted with EKHC leaders and technical staff working in EKHC' development programmes on their attitude towards the contributions of FBOs on improving ecological balance and food security, shows, however, mixed results. Participants seemed to have a balanced assessment on the work of FBOs in their areas of intervention. The respondents believed that FBOs' contribution in schooling and emergency relief are most effective and have the strongest positive impact. However, in total, they listed more weaknesses than strengths related to FBO services. The weaknesses ranged from tendencies to spiritualise every move they take to declining integrity. In conclusion the findings of the study show that the respondents are not too optimistic regarding the potential of the EKHC to contribute to ecological balance and food security. This is particularly sticking given the very ambitious goals of the EKHC strategic plan 2007–2012 as shown in Chapter 3.4.3.

There is a fairly high level of understanding among the EKHC staff on questions of ecological balance, food security and sustainable development as such. This underlines the strong and important role of the EKHC as a catalyst to mobilize rural communities to engage in actions that promote socio-economic and environmental transformation in Ethiopia. It is however, noteworthy that the interviewees gave rather 'apolitical' answers, and undervalued political, societal and historical reasons for food insecurity and environmental degradation. Issues like contested land use rights or political instability, for example, were rarely mentioned.

Furthermore there are indications that development related activities of the EKHC are still regarded as 'second class' services by some of its staff. There is, for example, a tendency to view the Community Development Major Programme at ETC with some degree of suspicion. Some people in the EKHC's inner circle continued to argue that FBOs should rather focus on 'spiritual' issues, leaving development education and the implementation of practical development projects – such as the dissemination of water filters and improved cooking stoves – for government development agents.

9.2. The Way Forward

This study has shown that FBOs have strong capacities and responsibilities not only to make noteworthy contributions to maintain and improve schooling systems in rural Ethiopia and to promote integrated approaches towards ecological balance and food security, but that they also have a huge potential to increase their impact on sustainable development in the future. In this regard, four points are proffered as a way forward, as shown in the following.

1. More recognition and integration
FBOs role as catalysts towards sustainable development particularly in remote rural areas of so-called developing countries is still often neglected by state and multilateral donors, research, media as well as other civil society and NGO actors. It is needed to strengthen the position of FBOs as important actors within the international development discourse. Better recognition of FBO's work through

open and critical evaluation and more adequate integration in higher-level development-related planning, such as national poverty strategies, is needed. While large scale FBOs such as EKHC have the capacities to promote significant socio-economic and – to a lower extent – environmental changes within communities, it needs far more concerted and coordinated actions beyond borders between faith-based and non-faith based NGO, or state and non-state actors.

2. *More empirical research*

It is absurd that development related work of FBOs is still absolutely under-represented in research and scientific development debates. Only very few studies have been conducted so far to empirically and critically document the strengths and limitations of FBOs in Ethiopia. The research study at hand is one contribution in this direction. However, given the pivotal position that both Christian and Muslim FBOs hold as local development brokers particularly in rural areas of Ethiopia, further research is definitely needed from different social-science disciplines to help understanding the potential, impact and limitations of FBOs in a development context.

3. *More investment in staff education and training*

FBOs must invest more in education and training of their own staff. The Community Development Major Programme of the EKHC can be seen as a promising step and showcase in this direction. However, more emphasis should be given to enable also village level staff of FBOs to undertake wholistic services in a way that they address socio-economic and environmental concerns in a problem-oriented and science-based manner.

4. *Upscaling local capacities*

Given their strong and deep-rooted local organisational structures of FBOs especially in rural areas of so-called developing countries, e. g. in the form or Bible schools and theological colleges, FBOs can upscale their contributions towards sustainable development by a number of activities. Local churches and Bible schools can do a lot more that organising Sunday services. Local churches have a high potential to act as wholistic sustainable development centres that can for example provide services such as childcare, family counseling, or advise for HIV / AIDS victims. Beyond local churches should invest more into mobilizing their members to take part in practical 'green' grassroots activities. One can, for instance, think about a "thousand-tree campaign", in which FBOs encourage each of their congregations to plant one thousand indigenous trees every year. If this would be done by the EKHC local schools alone it could lead in absolute terms to the planting of 6,000,000 new indigenous trees every twelve months!

References

Abebe Tadege (2008). Climate Change and Development Adaptation Measures in Ethiopia. Economic Focus: Bulletin of the Ethiopian Economics Association (EEA), Vol. 11, No. 1, December 2008

Abinet Gebrekidan (2006). Integrating Environmental Education into the Secondary and Senior Secondary Schools Curricula in Ethiopia, In Proceedings of the Conference of Teacher Education for Sustainable Development in Ethiopia, Organised by College of Education, Addis Ababa University, May 5 – 6, Debre Zeit

Abiy Hailu (2004). The Role of NGOs and Civil Society in Food Security. In Civil Society and Food Security: A Forum for Public Dialogue Organised by Forum for Social Studies and Ethiopian Economic Association, 24 February 2004, Addis Ababa (unpublished)

Ahm, P. (1998). Preparing a Sustainable Market for PV Solar Home Systems, in Shrestha, J. N. et al. (eds.), Role of Renewable Energy Technology for Rural Development, in Proceedings of International Conference on Role of Renewable Energy Technology for Rural Development, 12 – 14 October, 1998, Kathmandu, Nepal

Aklilu Dalelo (1998). Educators' Views about the Use and Protection of Natural Resources in Ethiopia: The Case of Teachers and School Administrators. The Ethiopian Journal of Education, Vol XVIII, No. 2, December 1998 p41 – 61

Aklilu Dalelo (2001). Natural Resource Degradation in Ethiopia: Assessment of Students' Awareness and Views: Flensburger Regionale Studien, Band 11

Aklilu Dalelo (2003). The Church and Socio-economic Transformation. Addis Ababa

Aklilu Dalelo (2006). Enabling Schools to Address Key Environmental Issues: Opportunities and Challenges. Journal of Education for Development, Vol. 1, No.1, September 2006, p37 – 56

Anderson, T. et al (2006). Trading with the Environment: Ecology, Economics, Institutions and Policy. London: EarthScan Publications Ltd.

Atkins, H. (2003). Stones of Remembrance (Compiled by Pual and Lila Balisky as part of the event to commemorate the 75th anniversary of the Ethiopian Kale Heywet Church)

Assefa Hailemariam (2003). Population Growth, Environment and Agriculture in Ethiopia, in Walta Information Centre (2003). Population and Development in Ethiopia: Now and in the Future, Symposium Proceedings, Addis Ababa, 17 June 2003

Bahru Zewde (1991). A History of Modern Ethiopia 1855 – 1974. London

Batchelor, P. (1993). People in Rural Development. The Paternoster Press

Belshaw, D. (2002). Accepted and Neglected Challenges in Christian Development Activity: Evidence from sub-Saharan Africa. Transformation 19/2 April 2002

Belshaw, D. (2006). Enhancing the Development Capabilities of Civil Society Organisations: With Particular Reference to Christian Faith-based Organisations (CFBOs). Transformation 23/3 July 2006

Bornstein, E. (2002). Developing Faith: Theologies of Economic Development in Zimbabwe. Journal of Religion in Africa, Volume 32.1 (2002)

Boulding, E. B. (1968). Beyond Economics: Essays on Society, Religion, and Ethics. The University of Michigan Press

Braaksma, D. (1994). Can Christian Development Work Fit on a Donkey's Back? Rethinking the Approach to Missionary Development Work among Muslim Pastoralists. Master of Theology Thesis Submitted to the University of Edinburgh

Bragg, W. G. (1987). From Development to Transformation. In Samuel, V. and Sugden, C. (eds.) (1987): The Church in Response to Human Need. Regnum Books

Cabraal, A. et al. (1996). Best Practices for Photovoltaic Household Electrification Programs: Lessons from Experiences in Selected Countries. World Bank Technical Paper Number 324. The World Bank, Washington, D.C.

Candland, C. (2001). Faith as Social Capital: Religion and Community Development in South Asia. Kluwer Academic Publishers

Cavalcanti, H. B. (2005). Food Security, In Dood, F. and Pippard, T. (2005), Human and Environmental Security: An Agenda for Change. Earthscan

Clarke, G. (2005). Faith Matters: Development and Complex World of Faith-Based Organisations. Paper Presented at the Annual Conference of the Development Studies Association, the Open University, Milton Keynes, 7–9 September 2005

Cunningham, W. P. and Cunningham, M. A. (2008). Principles of Environmental Science: Inquiry and Applications. New Delhi: McGraw-Hill Publishing Company Limited

Daniel W/Giorgis (2003). The History of EKHC Development Works with Focus on Kembata & Hadiya Church Development Program. A Paper Presented in the Workshop on "Future Direction of KHCDP" (unpublished)

Dercon, S. (1999). Ethiopia: Poverty Assessment Study. A Revised Version of a Report for IFAD

DeRose, L., Messer, E. and Millman, S. (1998): Who's Hungry? And How Do We Know? Food Shortage, Poverty and Deprivation. United Nations University Press

Dessalegn Rahmato (1988). Peasant Survival Strategies. In Penrose, A. (ed.), Beyond the Famine: An Examination of the Issues behind Famine in Ethiopia. Geneva

Dessalegn Rahmato (2007). Development Interventions in Wollaita, 1960s–2000s: A Critical Review. FSS Monograph No.4. Forum for Social Studies, Addis Ababa

Development Studies Associates (1998). Training on Food Security. Soddo and Hosanna

de Waal, A. (2000). Famine Crimes: Politics and the Disaster Relief Industry in Africa. Indiana University Press

Dicklitch, S. and Rice, H. (2004). The Mennonite Central Committee (MCC) and Faith-based NGO Aid to Africa. Development in Practice, Volume 14, Number 5, August 2004

Diesen, A. and Walker, K. (1999). The Changing Face of Aid to Ethiopia – Past Experience and Emerging Issues in British and EC Aid. www.christian-aid.org.uk

DPPC (2004). Ethiopia: National Information on Disaster Reduction: Report for the World Conference on Disaster Reduction (Kobe-Hyogo, Japan, 18–22 January 2005)

DTRC (Demographic Training and Research Centre), Addis Ababa University, and PSTC (Population Studies and Training Centre), Brown University (1998). Southern Nations, Nationalities and Peoples' Region Community and Family Survey 1997. Addis Ababa

ECA (Economic Commission for Africa) (1997). Sustainable Agriculture and Environmental Rehabilitation Program and the *Woreda* Agriculture and Rural Development Integrated Services: Main Report: Kembata Alaba Tembaro zone

EEA (Ethiopian Economic Association) (2005). Report on the Ethiopian Economy, Volume IV 2004/05. Addis Ababa

EEA (Ethiopian Economic Association) (2008). Report on the Ethiopian Economy, Volume VI 2006/07. Addis Ababa

ECFE (2005). ECFE Missions Research, in Partnership with Dawn Ministries. Addis Ababa

EKHC (2007). Ethiopian Kale Heywet Church Five Year Strategic Plan (2007 – 2012). Addis Ababa

Erango Ersado (2003). Strengths and Weaknesses of EKHC Development Work: The Case of Gedeo & Kembata and Hadiya zones. A Paper Presented in a Workshop on "Future Direction of KHCDP" (unpublished)

Ethiopian National Agency for UNESCO (2001). The Development of Education: National Report of Ethiopia (Final Version)

FAO (2008). Climate Change and Food Security: A Framework Document. Rome

Fassil G. Kiros (2005). Enough with Famines in Ethiopia: A Clarion Call. Addis Ababa

Fien, J. (2006). Teaching and Learning for a Sustainable Future: Unesco's New Multimedia Teacher Education Program, In Proceedings of an International Conference on Globalization and Education for Sustainable Development, 28 – 29 June 2005, Nagoya, Japan

Gatzweiler, F. (2007). Deforestation of Ethiopia's Afromontane Rainforests: Reasons for Concern. ZEF Policy Brief No. 7

Geest, W. V. (1993). Development and Other Religious Activities: Discussion Paper Prepared for Churches and Development Workshop June 14 – 15, 1993, Toronto, Canada

Getachew Diriba (1995). Economy at the Cross Roads: Famine and Food Security in Rural Ethiopia. Addis Ababa

Getnet Tamene (1998). Features of the Ethiopian Orthodox Church and the Clergy. Asian and African Studies, 7, 1998, 1, 87.104

Girma Kebede (1988). Cycles of Famine in a Country of Plenty: The Case of Ethiopia. GeoJournal 17.1, 125 – 132

Golly, F. B. (1999). A Premier for Environmental Literacy. University Press (India) Limited

Greste, P. (2006). Ethiopia's Food Addiction. www.bbc.co.uk/mpapps/pagetools/print/news-accessed on 07 August, 2006

Grootaert, C. and Bastelaer, T. V. (eds.) (2002). The Role of Social Capital in Development: An Empirical Assessment. Cambridge University Press

Goyder, H. and Wigbouldus, S. (2006). Ethiopia & Eritrea Appeal 2003 – 2006 Evaluation. Tearfund UK

GTF (Gudina Tumsa Foundation) (2003). Witness and Discipleship: Leadership of the Church in Multi-Ethnic Ethiopia in a Time of Revolution. The Essential Writings of Gudina Tumsa. Addis Ababa

Hailu, T. (1975). Ethiopia's Educational Development. Education in East Africa, Vol. 5. No. 2, 1975 p. 145 – 156

Haakansson, M. (2009). When the Rains Fail: Ethiopia's Struggle Against Climate Change. Kobenhaven: Informations Forlag

Harrison, P. (1990). The Greening of Africa: Breaking Through in the Battle for Land and Food. London

Horn, N. E. and Tefera Talore (2006). The EKHC Leadership and Ministry Transformation Project/Gilgal Baseline Survey: Findings of a Quantitative Study of Change in the Ethiopian Kale Heywet Church. Addis Ababa

Horne, R. E. and Frost, S. (1992). War, Famine and Environment in Eritrea. The International Journal of Environmental Education and Information, Vol. 11, No. 4, 293–306

Hughes, D. (1998). God of the Poor: A Biblical Vision of God's Present Rule. OM Publishing

Hussien Jemma (2001). The Debate over Rural Land Tenure Policy Options in Ethiopia: Review of the Post-1991 Contending Views. Ethiopian Journal of Development Research, Volume 23, Number 2, p. 35–84

IDCoF (International Development Consulting Firm) (2002). Food Security, Sustainable Agriculture and Trade: Issues of Concern for Lobby and Advocacy. A Report Prepared for the Christian Relief and Development Association. Addis Ababa

IPCC (2007). Climate Change 2007: The Physical Science Basis – Summary for Policymakers. Contribution of Working Group I to the Fourth Assessment Report of the Intergovernmental Panel on Climate Change.

Jansson, K. et al (1990). The Ethiopian Famine: Revised and Updated Edition. Zed Books Ltd.

Jialin, L. et al. (2009). Effects of Land Use Changes on Values of Ecosystem Functions on Coastal Plain of South Hangzhou Bay Bank, China. African Journal of Agricultural Research, Vol. 4 (5), p. 542–547, May 2009

Kayastha, Y. (2000). The Role of Solar Home Systems in the Promotion of Income Generating Activities in Selected Villages of Kavre District in Nepal. Unpublished MSc Thesis Presented to ARTES, University of Flensburg

Markakis, J. (1974). Ethiopia: Anatomy of a Traditional Polity. Clarendon Press

Marshall, K. (2005). Faith and Development: Rethinking Development Debates (http://web.worldbank.org

Marshall, K. and Saanen V. M. (2007). Development and Faith: Where Mind, Heart, and Soul Work Together. Washington: The World Bank

Melaku Girma (1994). An Investigation into the Integration of Environmental Education into Social Studies Course in Some Selected Teacher Training Institutes (TTIs) of Ethiopia. Unpublished MA Thesis Submitted to the School of Graduate Studies, Addis Ababa University, Addis Ababa

Mesfin Wolde-Mariam (1984). Rural Vulnerability to Famine in Ethiopia 1958–1977. Addis Ababa

Mesfin Wolde-Mariam (1991). Suffering under God's Environment: A Vertical Study of the Predicament of Peasants in North-Central Ethiopia. Berne

MoFED (Ministry of Finance and Economic Development) (2002). Ethiopia: Sustainable Development and Poverty Reduction Program. Addis Ababa

MoFED (Ministry of Finance and Economic Development) (2006). Ethiopia: Building on Progress: A Plan for Accelerated and Sustained Development to End Poverty (PASDEP), 2005/6–2009/10, Volume I: Main Text. Addis Ababa

MOLSA (Ministry of Labour and Social Affairs) (2002). Ethiopia's National Plan of Action for Children (2003–2010). Addis Ababa

Mulatu Bafa (2003). The Inception, Growth and Outcomes of the EKHCDP. A Paper Presented in the Workshop on "Future Direction of KHCDP" (unpublished)

Nigussie Zewdie (2003). The EKHC Development Work: Strengths & Weaknesses. A Paper Presented in a Workshop on "Future Direction of KHCDP" (unpublished)

Palmer, J. A. (1998). Environmental Education in the 21st Century: Theory, Practice, Progress and Promise. London

Pankhurst, R. (1966). Ethiopia, in Scanlon, D. G. (ed.), Church, State and Education in Africa. New York

Paterson, I. (2008). Final Report: Evaluation of SIM Canada's 2004–2007 Program (S62896), April 2008

Population Census Commission (2008). The 2007 Population and Housing Census Results of Ethiopia. Addis Ababa

Poyck, G. et al. (2009). Mid Term Evaluation of Basic Education Programme 2007–2010: Help a Child. Addis Ababa: Edburghconsultants

Purohit, S. S. and Agrawal, A. K. (2005). Ecology and Environmental Biology. Yodhpur: Shyam Printing Press

SNNPR Education Bureau (2001). Social Studies Student Text for Grade 8. Awassa

Shumete Gizaw (2006). Towards Effective and Efficient Environmental and Population Education for Sustainable Development in Ethiopia. In Proceedings of the Conference of Teacher Education for Sustainable Development in Ethiopia, Organised by College of Education, Addis Ababa University, May 5–6, Debre Zeit

Sirgiw Gelaw (2007). The Ethiopian Orthodox Church and Its Role in Development in 2020. Economic Focus Vol. 9, No. 5, June 2007

Stellmacher, T. (2007a). Governing the Ethiopian Coffee Forests. A Local Level Institutional Analysis in Kaffa and Bale Mountains. Bonner Studien zur Wirtschaftssoziologie. Bd. 27

Stellmacher, T. (2007b). The Historical Development of Local Forest Governance in Ethiopia – from Imperial Times to Military Regime of the *Derg*. Africa Spectrum 42 (2007) 3:519–530

Stellmacher, T. and P. Mollinga (2009). The Institutional Sphere of Coffee Forest Management in Ethiopia: Local Level Findings From Koma Forest, Kaffa Zone. International Journal of Social Forestry. 2(1): 43–66

Stellmacher, T. and R. Nolten (2010). Forest Resource Use and Local Decision Making in the Bale Mountains Coffee Forests, Ethiopia. In: I. Eguavoen & W. Laube (eds.): Negotiating Local Governance. Natural Resources Management at the Interface of Communities and the State, Lit Publishing, Berlin

Sultan, P. (2001). Church and Development: A Case Study from Pakistan. FACT Publications

Tassew Woldehanna (2004). New Pro-poor Policies, Pre-PRSP Experiences and the 2003–04 Budget in Relation to Agriculture and Food Security in Ethiopia. In Tassew Woldehanna and Walter Eberlei (eds.), Pro-poor Budgeting and the Role of Parliament in the Implementation of PRSP in Ethiopia. Addis Ababa

Tsele, M. (2001). The Role of the Christian Faith in Development. In Belshaw, D. et al.

(eds.), Faith in Development: Partnership between the World Bank and the Churches of Africa. Regnum Books International

Tyndale, W. (2001). World Faiths Development Dialogue. COMPAS Magazine, March 2001

WCED (World Commission on Environment and Development) (1987). Our Common Future. Oxford: Oxford University Press

Webb, P. et al. (1992). Famine in Ethiopia: Policy Implications of Coping Failure at National and Household Levels. International Food Policy Research Institute, Research Report 92

Workneh Negatu (2008). Food Security Strategy and Productive Safety Net Program in Ethiopia, in Taye Assefa (ed.), Digest of Ethiopia's National Policies, Strategies and Programs. Addis Ababa: Forum for Social Studies

World Evangelical Fellowship Theological Commission (1999). Evangelical Christianity and the Environment. In Samuel, V. and Sugden, C. (1999), Mission as Transformation: A Theology of the Whole Gospel. Regnum Books International

Wright, R. T. and Nebel, B. J. (2002). Environmental Science: Towards a Sustainable Future. New Jersey: Pearson Education

UNAIDS, Unicef and USAID (2003). Children on the Brink 2002: A Joint Report on Orphans Estimates and Program Strategies. New York

Unesco and the Government of Greece (1997). Education for Sustainable Future: A Transdisciplinary Vision for Concerted Action. A Background Paper for the International Conference on Environment and Society: Education and Public Awareness, Thessaloniki, Greece, 8–12, December

Unicef (2005). Excluded and Invisible: The State of the World's Children. New York

Appendix I

A questionnaire to be completed by church ministers (church elders, Evangelists, pastors, etc.) and church-based development workers.

Dear Respondent,

This questionnaire is meant to gather information for a study on the "Contribution of faith-based organisations to improving educational access, ecological balance and food security in Ethiopia", with particular focus on Evangelical churches. The EKHC has been selected for a detailed empirical study. The research is hoped to reveal, amongst other things, the strengths and weaknesses of faith-based organisations with regard to addressing the issues of educational development, environmental protection and food security in this country.

It is also hoped that the study will help to improve the quality of services currently offered by faith-based organisations (*FBOs*), thereby enabling them to properly play their role as 'light' and 'salt.' We thank you indeed for taking your precious time to contribute towards our study.

A. **General Information**

1. Sex: *Male*-------- *Female*--------
2. Name of your (EKHC) *Awraja*-------------------------------------
3. Your ministry at local church (if any)--------------------------------
4. Level of education (Please put an "X" mark on one of the following):
   Grade 12 or below-----------
   Certificate--------------------
   Diploma----------------------
   BA/BSc/BTh------------------
   MA/MS cot above------------
5. Profession---

B. **Please respond to the following 'general' questions in writing, as exhaustively as you can.**

1. How do you understand "development"? ---

   ----

2. How do you understand "poverty"? --

   ----

3. How do you see the relationship between environmental protection and food security?--------------
   ----

   ----

4. What do you think are the major causes of food insecurity in Ethiopia? (Please give only the three causes you think are most important).
   1. --
   2. --
   3. --

5. What do you think are the *unique* contributions of faith or religion to the socio-economic development of a given society?
   1. --
   2. --
   3. --

6. What do you think are the major strengths of faith-based organisations (FBOs) in bringing about socio-economic transformation *in Ethiopia*?
   1. --
   2. --
   3. --

7. What do you think are the major weaknesses of FBOs in Ethiopia in bringing about socio-economic transformation?
   1. --
   2. --
   3. --

8. What should FBOs in Ethiopia do to solve the problem of food insecurity in the country? (Please give only the three measures you think are most important).
   1. --
   2. --
   3. --

Appendix I

9. How do you rate the relationship Evangelical churches in Ethiopia have with governmental offices in the country?
   Excellent-------- Very good----------Good-------------- Poor----------

C. **Please respond to the following questions related to 'your local church' in writing, as exhaustively as you can.**

1. What are the *key values* your church wants to inculcate into its followers? (Please write down only the three values you think are most important):
   1. --
   2. --
   3. --
2. To what extent do you think ministers and members of your local church are committed to these values?
   Highly ---------- Moderately ----------Not much------
3. Does your local church currently undertake any social/development ministry (wholistic ministry)?
   Yes-----------------No------------------
4. If yes, please list down only three of the social/development ministries in your local church in order of relative strength
   1. --
   2. --
   3. --
   4. Who is responsible for coordinating the social/development ministries in your local church?
      ----
6. What strategies are used to get the resources (human, material, financial, etc.) for the social/development ministry? --
   ----
   ----
7. Does your local church get any assistance (technical, material, financial, etc.) from sources outside the church? Yes------------ No ------------------
   If yes, please describe the sources and the types of assistance -------------------------------------
   ----

   ----
8. To what extent is your church engaged in advocacy work?
   Highly ---------- Moderately ----------Not much------
   If there is any advocacy work, please give three examples below:
   1. --
   2. --
   3. --
9. Does its identity as a church make it easier or more difficult for *your local church* to do social/development work in your community? Easier----- More difficult-------
   In what ways? ---

D. Please indicate your views about each of the statements by putting a circle around one of the following alternatives: strongly agree (SA), agree (A), undecided (U), disagree (D), strongly disagree (SD).

s/n	Statements	Views				
1	The primary mission of a church is converting people to the Christian faith.	SA	A	U	D	SD
2	The church should not consider 'medical work' an opportunity to preach the Gospel.	SA	A	U	D	SD
3	In the Ethiopian situation, development cannot be achieved without external assistance.	SA	A	U	D	SD
4	Project staff or people supported by a project should not be directly engaged in Evangelism and church planting.	SA	A	U	D	SD
5	External assistance cannot bring about long-lasting community development.	SA	A	U	D	SD
6	Food insecurity in Ethiopia is more a problem of lack of enough food in the country than lack of access to food.	SA	A	U	D	SD
7	Genuine development should aim at change in spiritual circumstances as well as physical, social and political ones.	SA	A	U	D	SD
8	Evangelical churches in Ethiopia strongly preach about hard work, but do not practice it.	SA	A	U	D	SD
9	Traditional religions have a negative effect on the process of development.	SA	A	U	D	SD
10	Evangelicals in Ethiopia have a better reputation regarding hard work compared to followers of the Ethiopian Orthodox Church.	SA	A	U	D	SD
11	The ultimate outcome of spiritual intervention must be acceptance of the Christian faith.	SA	A	U	D	SD
12	Evangelicals in Ethiopia have a better reputation regarding hard work compared to the followers of Islam.	SA	A	U	D	SD
13	Science and technology can change the current state of destruction of the natural environment.	SA	A	U	D	SD
14	Community development work can be viewed as a pre-Evangelistic activity.	SA	A	U	D	SD
15	In Ethiopia, people who are engaged in spiritual ministry (e.g. pastors) are more respected than those engaged in other professions.	SA	A	U	D	SD
16	Evangelism and church planting should be an integral part of the development ministry.	SA	A	U	D	SD
17	Evangelicals in Ethiopia tend to discourage the accumulation of money and capital goods.	SA	A	U	D	SD
18	In church-based development work, socio-economic development should be the primary purpose.	SA	A	U	D	SD
19	Evangelicals in Ethiopia tend to discourage business and investment.	SA	A	U	D	SD
20	In church-based development work, spiritual transformation should be seen as a component of the socio-economic development process.	SA	A	U	D	SD
21	In church-based development work, an assessment of staff performance should also take efforts to bring people to a Christian faith into consideration.	SA	A	U	D	SD
22	Evangelicals in Ethiopia give equal value to work and worship.	SA	A	U	D	SD
23	All the activities underlying a wholistic development (health, education, agriculture, etc.) must be given equal emphasis to Evangelism and church planting.	SA	A	U	D	SD
24	In Ethiopia, spiritual leaders have more acceptance than political leaders in day-to-day activities like conflict resolution.	SA	A	U	D	SD
25	In Ethiopia, spiritual leaders have better opportunity to mobilise communities for development work than political leaders.	SA	A	U	D	SD
26	In church-based development work, there should be no interlinking between development activity (social, economic, health) and other religious activities (Evangelism and church planting).	SA	A	U	D	SD

Appendix I

27	Spiritual leaders are more respected in my own community than political leaders.	SA	A	U	D	SD
28	Development activities are causing massive destruction to the natural system.	SA	A	U	D	SD
29	In Ethiopia, FBOs have a key role to play in building the capacity of communities to attain food security.	SA	A	U	D	SD
30	Development and Evangelism should be regarded as distinct and separable tasks of the church.	SA	A	U	D	SD
31	The Ethiopian government recognises the contributions of FBOs in the country.	SA	A	U	D	SD
32	It is wrong to expect a genuine conversion 'to a given religion/faith as a means to benefit from development activities.'	SA	A	U	D	SD
33	How do you view the statement: "Give a man fish and you feed him for one meal, but if you teach him how to fish, you feed him for life"?	SA	A	U	D	SD
34	In church-based development work, more value should be attached to "life witness or presence ministry" than seeking to convert people.	SA	A	U	D	SD
35	Development activity, whether accompanied by specific spiritual changes or not, is important on its own merits.	SA	A	U	D	SD
36	One should not see education as a means of winning converts.	SA	A	U	D	SD
37	How do you assess the statement: "Nature must be saved or we humans will die"?	SA	A	U	D	SD